Introduction to

SHORE WILDFLOWERS

of California, Oregon, and Washington

REVISED EDITION

Philip A. Munz

Edited by Phyllis M. Faber and Dianne Lake

UNIVERSITY OF CALIFORNIA PRESS

Berkeley   Los Angeles   London

S0-CFO-418

California Natural History Guides No. 67

University of California Press
Berkeley and Los Angeles, California

University of California Press, Ltd.
London, England

© 2003 by the Regents of the University of California

Library of Congress Cataloging-in-Publication Data

Munz, Philip A. (Philip Alexander), 1892–1974
    Introduction to the shore wildflowers of California, Oregon, and Washington /
    by Philip A. Munz.
        p.   cm—California natural history guides ; 67.
    Includes bibliographical references (p.).
    ISBN 0-520-23638-6 (hc. : alk. paper).—ISBN 0-520-23639-4(pbk. : alk.
    paper)
    1. Wild flowers—Pacific Coast (U.S.)—Identification   2. Seashore plants—
    Pacific Coast (U.S.)—Identification.   3.— Wild flowers—Pacific Coast
    (U.S.)—Pictorial works.   4. Seashore plants—Pacific Coast (U.S.)—Pictorial
    works.   I. Title. II. Series.

    QK143 .M798   2003
    582.13´0979´09146—dc21

                                                                2002031964

Manufactured in China
12   11   10   09   08   07   06   05   04   03
10   9   8   7   6   5   4   3   2   1

The paper used in this publication meets the minimum requirements of
ANSI/NISO Z39.48-1992 (R 1997) (*Permanence of Paper*). ♾

CALIFORNIA NATURAL HISTORY GUIDES

INTRODUCTION TO SHORE WILDFLOWERS OF CALIFORNIA, OREGON, AND WASHINGTON

California Natural History Guides

Phyllis M. Faber and Bruce M. Pavlik, General Editors

The publisher gratefully acknowledges the generous
contributions to this book provided by

the Moore Family Foundation
Richard & Rhoda Goldman Fund
and
the General Endowment Fund of the
University of California Press Associates.

———————

Grateful acknowledgment is also made to
John Game and William T. and Wilma Follette
and
to the California Academy of Sciences

CALIFORNIA
ACADEMY OF
SCIENCES

for their contribution of photographs.

CONTENTS

EDITOR'S PREFACE
TO THE NEW EDITION

Shore Wildflowers Of California, Oregon, and Washington has introduced thousands to the wildflowers of the coastal areas of California. Since it was first published in 1964, a number of plant names have been changed, and, in some cases, new information has been obtained. In this revised and updated edition, a number of steps have been taken to make the book current in content and appearance.

New plants have been added to give a more representative distribution, north to south. Dr. Robert Ornduff wrote descriptions for these plants before his untimely death in 2000. He also wrote an introduction to coastal habitats for this edition of the shore wildflower book as well as for the other three Munz wildflower books, which are also being revised. The new *Mountain Wildflowers* debuts with this volume, and *Desert Wildflowers* and *Spring Wildflowers* will follow in 2004.

Scientific names for each plant have been made to conform to the current California authority, the *Jepson Manual: Higher Plants of California,* J. Hickman, editor (University of California Press, 1993). In addition, each plant in this edition has been given a common name using the following sources, listed here in descending order of preference: the *Jepson Manual;* P. Munz, *California Flora* (University of California Press, 1959); and L. Abrams, *Illustrated Flora of the Pacific States* (Stanford University Press, 1923–60). As before, common names follow established convention for hyphenation: if a plant's common name indicates a different genus or family, a hyphen is inserted to show that the plant does not actually belong to that genus or family. Thus, "skunk-cabbage" is hyphenated because the plant it refers to is not in the cabbage genus nor the cabbage family, but "tiger lily" is not hyphenated because the plant it refers to is in the lily genus, as well as the lily family.

Each plant description has been carefully checked for accuracy and currency. In several cases, taking into account research done in the last 50 years, a description that applied to an entire species in the first edition perhaps only pertains to a variety or subspecies, or vice versa, today. Some species have been absorbed into other species, and some have been split into varieties or subspecies. Some varieties or subspecies have even become separate species.

Dianne Lake has brought the scientific names up to date and has made appropriate revisions to the 1964 plant descriptions. The Press is grateful to her for her meticulous work. Many of the lively drawings of Jeanne Janish, mentioned in Munz's introduction, have been retained. New color illustrations and new design features have been added to make the book more user friendly. The Press is especially grateful to the team of Wilma and William T. Follette for providing so many of the photographs and for their donation to the Press.

Many of the plants found in this book have been severely reduced in their range by coastal development and by invasive weeds. Dr. Munz's intent was to include a variety of plants, both those commonly found, as well as those less frequently encountered but of special interest. We have removed species where only a few populations remain. A few plants in this book are rare in California but are more common in Oregon, Washington, or in Baja California. Users of this book are urged to respect all native plants and refrain from picking or collecting specimens. Enjoy our unique flora but leave it to flourish for future generations.

Phyllis M. Faber
September 2002

ACKNOWLEDGMENTS

Most of the drawings for this book were made by Jeanne R. Janish, whose illustrations in books on western botany are so well known and so successful in recreating the appearance of a plant in three dimensions from a pressed specimen that any botanical author feels proud to say that Mrs. Janish is the illustrator. Her drawings in this book are always indicated by the letter "J."

Once again, I am gratefully indebted to Susan J. Haverstick, who has so understandingly edited the previous books: *A California Flora, California Spring Wildflowers, California Desert Wildflowers,* and *California Mountain Wildflowers.* Her suggestions on the text itself and her ideas have been most helpful. I thank her once more for participating in this final wildflower book on California, Oregon, and Washington.

Philip A. Munz
February 1964

This book is the fourth in the series of so-called wildflower books that cover California plants, but in this case, a wider geographical range is represented, namely, all three Pacific states. The original and main purpose of these books is to enable general readers to identify plants they find growing in the wild. This book presents how to recognize these plants, what they are called, where you can find them, and what their general relationship is to other plants.

I have always thought that many may wish to know not only the pretty and conspicuous plants, but other plants as well. In our coastal salt marshes, many different plants grow that are striking by virtue of peculiar structure: fleshiness, jointed stems, odd-looking inflorescences, and so on. On coastal dunes, root parasites can be found, which lack chlorophyll and are not showy but display intriguing, small, purple flowers with white rims. You may want to know the differences among grasses (Poaceae), sedges (Cyeraceae), and rushes (Juncaceae), which are all more or less alike in their small flowers and are often conspicuous as dune binders and marsh inhabitants. Because the number of kinds of trees that come down to the shore and its adjacent bluffs is limited, it seems appropriate to include as many of them as space permits. So, although some of my reviewers in the past have criticized the selection of species used for my wildflower books and the space given to inconspicuous species, I tried to consider all the species that might arouse your curiosity.

Although I have selected the areas covered by the four volumes (*California Spring Wildflowers, California Desert Wildflowers, California Mountain Wildflowers,* and *Shore Wildflowers of California, Oregon, and Washington*) in such fashion as to have quite different species represented in the four books, a small amount of duplication is inevitable; however, only a few species are repeated.

Shore Climate

As explained in the first paragraph, the species represented in this book are not limited to the state of California. But as you study beach and shore species, you will immediately see that many grow far to the north, even to Alaska. The cooler, more regular, climate of the coast as compared with that of the interior and the longer-lasting effect of even small amounts of rain along the coast as compared with inland valleys make it obvious that the same plant may range, if not from southern California, at least from central California far into Oregon or Washington or even to British Columbia or Alaska. Coastal summers are relatively cool, and coastal winters are relatively warm. Hence, this book attempts to help you identify not only California shore wildflowers, but those of Oregon and Washington. In fact, some species are included from these two latter states that do not range far southward into California or only range into its extreme northern part.

What Is the Shore?

The first question that arose in my mind when I was asked to write a book on shore wildflowers was what to include. Naturally, the actual sandy beach and dunes, which together might be called the coastal strand, would come first. Next would be the coastal salt marshes, with their rather distinctive flora. Then I would include the bluffs along the coast, especially as far inland as salt spray seems to influence. Such influence is evident in southern California, particularly in the appearance of desert plants, such as bladderpod *(Isomeris arborea)*, or of more or less saline conditions. But as you go north, this influence decreases, although in much of central California there is a well-marked zone of coastal scrub, which has quite different assemblages of plants than in the redwood or other forests behind it. I assume that this coastal scrub is still due, at least in part, to the influence of salt spray. But still farther north with still greater rainfall, as in extreme northern California and in Oregon and Washington, the forests come right down to the bluffs and to the actual sand of the back beaches. I believe that with the high rainfall in this area, any possible effect of salt is almost immediately leached out and that the actual shore is invaded by plants such as pearly everlasting *(Anaphalis margaritacea)* and giant horsetail *(Equisetum telmateia)*, normally of wooded places, whereas a little to the south, species such as false lily-of-the-valley *(Maianthemum dilatatum)* come out on to the actual beach only along freshwater streams. In some ways, then, the northern coast has more species that normally grow in the adjacent forests than does the southern. Then, too, with the greater rainfall in the north, sandy areas are more easily taken over by cordgrass *(Spartina foliosa)* and other perennials, and there is not the development of as rich a strand flora, for the most part, as there is in Monterey and San Luis Obispo Counties of California, although a possible exception might be cited at Gold Beach, Oregon.

Characteristics of Shore Plants

Apparently the most important single factor in the environment of shore plants that sets them apart from those farther inland is the presence of salt or salts in the soil from seawater. Dissolved salts mean physiological dryness for the plant, which then has to contain within itself a higher percentage of dissolved substances to pull in water by osmosis than it would if in pure water. This is true whether growing in arid regions such as the desert, where the dissolved salts in the soil may be appreciable and where they may even coat the surface with a layer of so-called alkali, or whether found along the sea coast.

Usually plants of these two types of environments have a

reduced evaporating surface as compared with those in a mesophytic environment, which has an abundance of good water in the soil, as in the garden or in a region of high rainfall. Reduction of evaporating surface may cause the development of thickened, fleshy leaves or the replacement of functional leaves by fleshy green stems with reduced evaporative surfaces. In either case, there is a resultant succulence, which you will notice if you are acquainted with the plants of an inland environment and are at the beach. You may run into species closely related to those that are familiar, but quite different in their succulence and compactness, or you may find more succulent forms of the same species. Examples are the sand-verbenas (*Abronia* spp.) and fiddlenecks (*Amsinckia* spp.) of the coast and of the interior.

Halophytes are plants specially adapted to life in soils with high concentration of salts. Good examples are some of the species of saltbush (*Atriplex* spp.), saltgrass (*Distichlis* spp.), pickleweed (*Salicornia* spp.), sea-blite (*Suaeda* spp.), and sea-fig and Hottentot-fig (*Carpobrotus* spp.). Some of these are found on sandy strands, others are found in the coastal salt marshes. For the most part these halophytes are not beautiful, but they can be quite striking in appearance.

Which Wildflowers Are in This Book?

I have used the term "wildflower" very loosely, as mentioned previously, making it almost synonymous with the word "plant," or perhaps better, "higher plant." Two flowering plants that grow entirely submerged, especially in shallow bays, are eel-grass (*Zostera* spp.) and surf-grass (*Phyllospadix* spp.). These plants are often cast upon beaches with pieces of kelp or seaweed. I also dedicate a small section to some of the coastal ferns and horsetails, which do not produce flowers. And we do not think of shrubs and trees as wildflowers, but several of them are discussed. In other words, I have attempted to include the more interesting higher plants found near or on the shore, as well as the showy forms such as azaleas (*Rhododendron* spp.), California poppy *(Eschscholzia californica)*, and foxglove *(Digitalis purpurea)*.

Of course, the scope of a small book like this is limited and it cannot possibly contain all the plants found in the area under discussion. Often, this book will indicate that the species you identify is a *Gilia*, a *Phacelia*, or a *Castilleja*, but perhaps not the species actually illustrated. When more information is needed than is available in this book, refer to the larger, more complete volume *The Jepson Manual: Higher Plants of California*, edited by J. Hickman (University of California Press, 1993), which is available in California stores and libraries. Also very useful, although some of the plant names are now different, is *A California Flora and Supplement*, by P. Munz and D. Keck (University of California Press, 1973).

For plants found farther north, see *Flora of the Pacific Northwest*, by C. L. Hitchcock and A. Cronquist (University of Washington Press, 1973); *Illustrated Flora of the Pacific States*, by L. Abrams (Stanford University Press); or *A Manual of the Higher Plants of Oregon*, by M. E. Peck (Oregon State University Press, 1961). Be aware, however, that many of the names in the latter books will be out of date.

How to Identify a Wildflower

It is impossible to discuss plants and their flowers without using the names of their parts. But only the most necessary terms have been included, some of which are defined here. Consult the glossary for other terms that are unfamiliar to you. In the typical flower we begin at the outside with the sepals, which are usually green, although they may be of other colors. The sepals together constitute the calyx. Next comes the corolla, which is made up of separate petals or petals

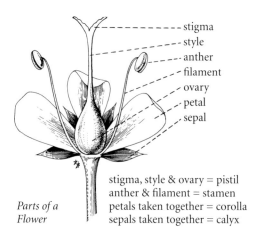

Parts of a
Flower

stigma, style & ovary = pistil
anther & filament = stamen
petals taken together = corolla
sepals taken together = calyx

A representative flower

grown together to form a tubular, bell-shaped, or wheel-shaped corolla. Usually the corolla is the conspicuous part of the flower, but it may be reduced or be lacking altogether (as in grasses and sedges), and its function of attraction of insects and other pollinators may be assumed by the calyx. The calyx and corolla together are sometimes called the perianth, particularly where they are more or less alike. Next, as we proceed inward into the flower, we usually find the stamens, each consisting of an elongate filament and a terminal anther where pollen is formed. At the center of the flower is one or more pistil, each with a basal ovary containing the ovules, or immature seeds, a more or less elongate style, and a terminal stigma with a rough, sticky surface for catching pollen. In some species, stamens and pistil are borne in separate flowers or even on separate plants. In the long evolutionary process by which plants have developed into the many diverse types of the present day and by which they have been adapted to various pollinating agents, their flowers have undergone very great modifications, and so now we find more variation in the flower than in other plant parts. Hence, plant classification is largely dependent on the flower.

To help you identify a flower, either a photograph or a drawing is given for every species discussed in detail, and the flowers are grouped by color. In attempting to arrange plants by flower color, however, it is difficult to place a given species to the satisfaction of everyone. The range of color may vary so completely from deep red to purple, from white to whitish to pinkish, or from blue to lavender that it is impossible to satisfy the writer himself, let alone the readers. I have done my best to recognize the general impression given with regard to color and to classify the plant accordingly, especially when the flowers are minute and the general color effect may be caused by parts other than the petals. My hope is that by comparing a given wildflower with the illustration it resembles within the color section you think is most correct and then checking the facts given in the text, you may, in most cases, succeed in identifying the plant.

Philip A. Munz
Rancho Santa Ana Botanic Garden
Claremont, California
February 10, 1964

INTRODUCTION TO COASTAL PLANT COMMUNITIES

by Robert Ornduff

Coastal areas of the Pacific states are popular destinations for vacationers, who are attracted by the broad sandy beaches, the surf, grassy coastal prairies, marshes, and coniferous forests. Visitors come to these areas to picnic, swim, surf, sunbathe, beachcomb, hike, fish, dig clams, observe birds, cool off, or simply to relax and enjoy the spectacular landscapes of our Pacific coastline. Many visitors are attracted to the beauty of the wildflowers, shrubs, and trees that are so richly represented along the coast; these can be seen at almost any season of the year. This guide introduces you to some of the common wildflowers, shrubs, and trees that occur in diverse coastal habitats of Washington, Oregon, and California. These habitats include stretches of sandy beaches and dunes, saltwater and freshwater marshes, coastal prairies, shrublands, riparian woodlands, and coniferous forests.

Coastal Habitats

The Pacific coastline of Washington, Oregon, and California is well over 1,000 miles long. It is marked by a diverse topography and bays ranging in size from Washington's enormous Puget Sound to the very small Bolinas Lagoon in

Marin County, California. Salt marshes are present on the margins of many of these bays and are important nesting sites for birds, resting and feeding sites for migratory birds, and spawning sites for marine invertebrates and fish. From the British Columbian border to Mexico's Baja California Norte, the backdrop of the coastline is formed by mountains that range in stature from the lofty and spectacular Olympic Mountains and northern Cascade Range of Washington to the lower and older coast ranges of Oregon and California. All along this coast, windswept promontories—capes, points, or heads—extend out into the ocean and support distinctive plant life. Offshore islands are common along the coast; most of these islands are small and virtually inaccessible to humans. These islets are important nesting and roosting grounds for sea birds and calving grounds for marine mammals. Many of the larger ones, such as the Channel Islands of southern California, are parks and reserves and receive large numbers of visitors each year. The islands in Puget Sound are home to several thousand residents and many summer visitors.

Along parts of the Pacific Coast, the ocean sweeps up onto sandy beaches, but in other places it pounds directly against rocky cliffs and headlands that are interspersed with these beaches. In some areas, large dune systems extend a few miles inland from the immediate coast. Some of these dunes are in nearly constant motion because of wind and pose threats to established vegetation, road systems, and houses and other buildings in their paths. Many coastal coniferous forest communities are limited to areas under a maritime influence, and in places, these forests occur very near the mean high-tide line. These forests include the Sitka spruce forests of the Pacific Northwest, the redwood forests that extend from southwestern Oregon to central California, and the Torrey pine forests of extreme southern California. All along the Pacific Coast, there are flat or rolling lands that are (or were) dominated by perennial grasses and herbs; these are coastal prairies, which are often carpeted with showy wildflowers in late spring and early summer. Patches of shrublands are frequent, but their component shrubs may also form the understory of the coastal forests. Freshwater marshes and ponds are common along the coast, especially from central California northward. The banks of rivers and streams that empty into the sea are lined by riparian woodlands.

Many coastal areas are now occupied by agricultural lands or urban areas and lack the native plants and animals that once lived there. Fortunately, however, many scenic portions of the coast are protected in national, state, and county parks, nature reserves, and recreation areas where the native biota can still be seen. Some reserves have been established primarily to protect the unusual plants that grow in them. These range in size from Redwood National Park in northern California to Darlingtonia Wayside in Oregon (named after the insectivorous California pitcher plant [*Darlingtonia californica*]).

Coastal Plant Communities

A plant community is an aggregation of plant species and other organisms that live together, interact with each other, and adapt to a specific set of environmental conditions. The nature of interaction among species may be subtle, but the environmental conditions that determine which plant community is present on a site may be striking. Wherever a given set of environmental conditions occurs, one expects to find the plant community that is adapted to these conditions as well. For example, coastal beach and dune plants occur only on unstable beach sand, coastal salt marshes occur in areas influenced by tidal action, riparian woodlands occur on the banks of freshwater streams and rivers, and coastal coniferous forests usually occur on deep, fertile, relatively moist soils.

The plants that are described in this book all occur on or near the coast and live in a maritime climate. This climate is characterized by relatively mild summers and winters and persistent summer fogs that serve to moderate temperatures and reduce water loss from plants and soil. Mean annual precipitation ranges from 120 inches on parts of the northern Oregon and northern Washington coasts to 16 inches along parts of the California coast from Monterey Bay southward. The mean minimum winter temperatures are above freezing along the entire coastline. Areas on the Washington and Oregon coasts experience occasional winter frosts, whereas frost is virtually unknown along much of the central and southern California coast. Plants along much of the immediate coast must endure the drying effects of strong, salt-laden onshore winds.

Most of the plants described in this book are native to the region, but some are introductions from other countries and are now well established. As you go from north to south along the Pacific Coast, you will see perceptible changes in plant distributions. A few plant species occur nearly throughout the area covered by this book; these include miner's-lettuce *(Claytonia perfoliata)*, silver beachweed *(Ambrosia chamissonis)*, bracken fern *(Pteridium aquilinum)*, and some introductions such as field mustard *(Brassica rapa)*. Others, such as yellow sand-verbena *(Abronia latifolia)*, salal *(Gaultheria shallon)*, and oceanspray *(Holodiscus discolor)* occur in most of the region but not in extreme southern California. Some common coastal species of the Pacific Northwest do not extend into California. Other species are common along the California coast and reach their northern limit on the southern or central Oregon coast. A few species included in this book are very limited in distribution.

Coastal Beaches and Dunes

Sandy beaches and dunes are common but discontinuous all along the Pacific Coast. In the Pacific Northwest, the most extensive dune system is near Coos Bay, Oregon; in California, there are notable dune systems near Eureka, Point Reyes, Monterey Bay, and in southern San Luis Obispo County. Although beaches and dunes may be appealing to human visitors, they are harsh environments for plants. Strong winds blow the unstable sand about, often exposing root systems or burying entire plants. These winds carry abrasive sand grains and deposit salt spray on the foliage. Sand has a poor water-holding capacity: a prolonged heavy rain may wet the soil less than an inch below the surface, and once saturated with water, sand dries quickly. Sand is usually deficient in the nutrients needed for normal plant development. Humus is absent. On a warm summer day, the sand surface may reach temperatures of 150 degrees F—hot enough to blister the sole of a human foot.

Coastal beaches and dunes are often characterized by a low plant density—less than 20 percent of the surface area may have a plant cover. Species diversity on a site may be low; sometimes only a half-dozen species occur over large areas. Plants of these habitats typically are perennial, have prostrate, creeping stems that often produce roots along their length, can reproduce by vegetative means as well as by seeds, and have succulent leaves, grayish leaves, or both. These features are all related to the instability of the sandy substrate, strong winds, poor water-holding capacity, and high surface temperatures on a summer day. Despite the harshness of this environment, many plants that grow here have showy flowers. The common names of many of the plants in these habitats reflect their narrow ecological preferences: beach morning glory *(Calystegia soldanella)*, beach-primrose *(Camissonia cheiranthifolia)*, silver beachweed *(Ambrosia chamissonis)*, dune buckwheat *(Eriogonum parvifolium)*, and dune tansy *(Tana-cetum camphoratum)*. Because of the weak seasonality of these coastal habitats, plants can be found in flower nearly every month of the year. A visit to most Pacific Coast beaches on New Year's Day usually reveals a surprising number of plants with a few flowers.

Despite these harsh conditions, or perhaps because of them, coastal beaches and dunes are fragile ecosystems. Many have been severely disturbed by human trampling, grazing animals, off-road vehicles, and the invasion of alien plant species. Some of the latter, such as the South African Hottentot-fig *(Carpobrotus edulis)* and European beachgrass *(Ammophila arenaria)* were introduced intentionally to stabilize the shifting sands. Both species may form large, dense patches that have eliminated nearly all the native plants formerly on the site. Impressive but expensive restoration programs using only native species have been successfully carried out in a few areas along the coast, particularly in California.

Coastal Prairies

A prairie is a plant community dominated by perennial grasses. Coastal prairies occur sporadically along the entire Pacific Coast and are particularly frequent from southwestern Oregon to the Monterey Bay area. Although this community is dominated by grasses, it produces displays of colorful wildflowers in spring and early summer that rival those of interior regions. These include baby blue-eyes *(Nemophila menziesii)*, blue-eyed-grass *(Sisyrinchium bellum)*, various buttercups *(Ranunculus* spp.*)*, California poppy *(Eschscholzia californica)*, checker mallow *(Sidalcea malviflora)*, and goldfields *(Lasthenia* spp.*)*. Species diversity is fairly high, especially compared with the often adjacent sand and dune communities that are species poor. Coastal prairies often occur between beaches and dunes and coniferous forests. They may form a mosaic with coastal scrub or forest due to local soil conditions or perhaps fire history. Where there are gaps in coastal mountains, coastal prairie species grow intermingled with those typical of inland prairies. Coastal prairies occur up to 3,000 feet in elevation and 60 miles or so inland where maritime influences extend far from the coast. An interesting example of an alpine coastal prairie is on Saddle Mountain, just behind Seaside, Oregon, and easily accessible by a trail. The balds on the hills of the coast redwood belt of northern California likewise are interior coastal prairies.

Coastal prairies experience a moderate climate, strong winds, and prolonged fogs. Unlike the sand of adjacent beaches and dunes, soils underlying coastal prairies are stable and rich, have a high clay and humus content, and a high water-holding capacity. These soils are the type a gardener would treasure in the backyard. These horticulturally desirable qualities have not gone unnoticed, and many areas of coastal prairie have been transformed into dairy ranches whose pastures have been seeded with alien grasses, or planted with artichokes, fava beans, brussels sprouts, and various plants grown for cut flowers.

Native grasses of undisturbed coastal prairies form a dense, evergreen turf. Trees are absent or may occur as scattered individuals. If shrubs are present, they are either prostrate, scattered, or both. These prairies are subject to intense grazing pressure by herbivores such as insects, rodents, rabbits, deer, elk, and domestic animals. They are subject to periodic fires, but the role of fire in shaping this plant community is not well understood. In places, pocket gophers (*Thomomys* spp.) keep the soil in a constant state of disturbance. In some areas, the floral display by spring annuals and summer perennials is spectacular: there are masses of goldfields (*Lasthenia* spp.), large patches of iris (*Iris* spp.), blue or yellow lupines (*Lupinus* spp.), red or orange paintbrushes (*Castilleja* spp.), and pink checker mallow *(Sidalcea malviflora)*. Some observers believe that these floral displays are more impressive where there is some level of grazing, but this issue remains a matter of controversy.

Coastal Salt Marshes

The margins of coastal bays, inlets, estuaries, lagoons, and riverbanks under tidal influence and with saline or brackish water are frequently occupied by coastal salt marshes. These marshes occupy the upper intertidal zone from mean high-tide level to extreme high-tide level. They are dominated by rather few species of evergreen herbaceous plants, most of which are low in stature, form a dense vegetative cover, offer a monochrome coloration, and have inconspicuous flowers. At any one site the species diversity is usually very low, with perhaps only six or so plant species present in a given salt marsh. Although some coastal salt marsh species have fairly restricted distributions, others occur along the entire Pacific Coast.

Coastal salt marshes are flat but gently slope upward from open water toward dry land. Their soils are saline to various degrees, and where they are more or less continually moist they usually have very poor oxygen levels (the fetid odor emitted by some salt marsh soils results from the metabolic activities of microorganisms that do not require free oxygen to thrive). The striking zonation in the distribution of the plants that grow in these marshes is related to tidal action, with some species growing only where they are inundated daily by the tides (such as California cord grass *[Spartina foliosa]*) and others growing on higher ground where inundation occurs only during very high tides or during winter storms (such as saltgrass *[Distichlis spicata]*). The differences in frequency and length of inundation result in gradations of soil salinity and soil moisture levels.

Coastal salt marsh plants are halophytes, plants that grow in saline soils. Some halophytic species can exclude salts (mostly sodium chloride) from entering their tissues, but others accumulate and sequester salts within their cells where they do not interfere with metabolism. The tissues of accumulator species such as pickleweed thus have a salty taste. Other salt marsh plants take up salts via their roots and excrete them via their leaves; tiny salt crystals can be seen on their leaf surfaces. Some species that grow in areas with a low soil-oxygen content have hollow tubes in their stems and leaves that transport oxygen downward from leaves to the root systems. Many salt marsh species such as saltgrass *(Distichlis spicata)* and pickleweeds (*Salicornia* spp.) are rhizomatous and form enormous clones via vegetative growth. The flowers of salt marsh plants typically are inconspicuous and often appear very late in the season. Midsummer visitors often notice large tangles of bright orange threads that appear in some salt marshes; these are the stems of a leafless plant, salt marsh dodder *(Cuscuta salina),* that parasitizes other salt marsh plants.

Despite their generally drab appearance, low diversity of plant species, and low vegetation, coastal salt marshes are extremely productive, that is, the levels of photosynthesis, or plant productivity, per unit area are very high. These marshes harbor large numbers of algae and other small marine organisms and are important feeding and nesting areas for resident and migratory waterfowl. Because these marshes are often near harbors and flat and can be kept dry by levees, they have disappeared or have become sharply reduced in extent from agriculture and other forms of development usage along many parts of the Pacific Coast. In California, more than 80 percent of the area occupied by coastal salt marshes in the early 1900s has been altered so drastically that these marshes have disappeared. In some areas, such as around San Francisco Bay, less than five percent of the former salt marshes persist.

Coastal Scrub

Coastal scrub refers to the shrub-dominated plant communities that occur along the entire Pacific Coast. These communities are more extensively developed in California and in southern Oregon than they are to the north. In Washington, Oregon, and northern California, coastal coniferous forests often have an understory of coastal shrub species that grow without a forest canopy farther to the south or to the coastward side of these forests. The shrubs in coastal scrub generally form a single layer, that is, mature individuals of different species are approximately the same height.

Coastal scrub is floristically diverse. Although some shrub species such as coyote brush occur almost throughout the range of this plant community, other species occur only regionally. Thus the species composition of southern Oregon coastal scrub is quite different from that in southern California. Evergreen or winter-deciduous shrubs are common in the north, whereas drought-deciduous and succulent species (including cacti) are more common in the south. These regional differences in plant behavior are likely the result of a wetter climate in northern areas than in southern ones.

Coastal scrub occurs intermixed with coastal prairie or as a fringe along the edge of coastal coniferous forests. It often occupies soils that are too poor, exposed, or unstable to support coastal prairie or coniferous forest. In southern California, coastal scrub extends as far as 60 miles inland from the coast, apparently in response to deep inland incursions of maritime influences (especially summer fog) in this region. In some coastal areas, after a forest is destroyed by fire, the most abundant woody species to appear after the fire are shrubs. These sites are temporarily converted from forest to scrub. After several years, conifer seedlings become established and eventually overtop and shade out many or most of the shrubs.

Freshwater Habitats

Where rivers and streams traverse coastal mountains and find their way to the Pacific, they are lined with a distinctive array of tree, shrub, and herb species that are limited to their banks. Red alder *(Alnus rubra)* is a common, deciduous riparian tree festooned with tiny, dry, female cones when leafless in winter. Ferns, horsetails (*Equisetum* spp.), box elder *(Acer negundo),* willows (*Salix* spp.), western azalea *(Rhododendron occidentale),* sedges (*Carex* spp.), and a variety of other plants occur in such areas as well.

Freshwater marshes and seeps are worth visiting to see the unusual plants that grow only in these habitats. These wet habitats in northern California, Oregon, and Washington are brightened in early spring by the large, bright yellow inflorescences of yellow skunk-cabbage *(Lysichiton americanum),* so called because its sweet fragrance is considered skunklike by some. After the flowers fade, the gigantic leaves lend a tropical look to these swamps. Yellow monkeyflowers (*Mimulus* spp.) abound in other spots, flowering over much of the year. Although not as showy as its tropical relatives, the flowers of the stream orchid *(Epipactis gigantea)* are intricate miniatures.

Coastal Forests

Forest communities occur along the Pacific Coast more or less continuously from northern Washington to central California. South of there, coastal forests and woodlands are discontinuous. The species of trees that occur in these forests differ as you go from north to south. In Oregon and Washington, these forests are dominated by conifers such as Sitka spruce *(Picea sitchensis),* western hemlock *(Tsuga heterophylla),* western red-cedar *(Thuja plicata),* and Douglas-fir *(Pseudotsuga menziesii).* Some of these trees commonly grow in pure stands, but others typically grow with other conifers or hardwoods. In southwestern Oregon, redwood trees *(Sequoia sempervirens)* begin to appear and grow in nearly pure stands southward to southern Monterey County. On drier substrates with reduced nutrient levels in coastal California, bishop pine *(Pinus muricata)* and Monterey pine *(P. radiata)* predominate. Monterey cypress grows naturally only on the Monterey Peninsula. Torrey pine *(P. torreyana)* grows in San Diego County and on Santa Rosa Island. Locally along the coast from southwestern Oregon southward there are forests composed largely of California bay *(Umbellularia californica)* (called Oregon myrtle in Oregon), and elsewhere one finds various other trees such as coast live oak *(Quercus agrifolia)*, tan-oak *(Lithocarpus densiflorus)*, and big-leaf maple *(Acer macrophyllum)* growing together and with various conifers.

The coastal forests of the Pacific Coast occur in a region that is relatively well watered (except for southern California) and has mild temperatures year round. The soils generally are deep, moist, and rich. Summer fogs help keep summer temperatures moderate and the soil moisture levels high. Fog condenses on the foliage of the trees and drips off, in places adding 10 or more inches to the effective annual rainfall beneath the trees. Many of these coastal conifers commonly reach ages of 1,000 years or more. The tallest trees in the world occur on the Pacific Coast, and the tallest one is a redwood *(Sequoia sempervirens)* in northern California that is over 350 feet tall. However, a Douglas-fir *(Pseudotsuga menziesii)* in Olympic National Park, Washington, is 326 feet tall, and a Sitka spruce *(Picea sitchensis)* in the same park is 305 feet tall.

Often, a single conifer species is dominant on a site. This is due to differences among the conifer species in requirements for moisture, soil characteristics, tolerance of flooding, fire history of the area, and other factors. Because coastal coniferous forests are dense, it is often very dark beneath the trees and refreshingly cool on warm summer days. The understory may consist of a dense layer of shrubs, most species of which also occur nearby in scrub communities without a forest canopy (e.g., salal *[Gaultheria shallon]*, California huckleberry *[Vaccinium ovatum]*, salmonberry *[Rubus spectabilis]*, poison-oak *[Toxicodendron diversilobum]*). Where the forests are old and dense, the forest floor receives little light and tree roots compete for moisture and nutrients. Here, there are few or no shrubs, and the forest floor is occupied by a few species of herbaceous plants such as the handsome evergreen western sword fern *(Polystichum munitum)*, redwood-sorrel *(Oxalis oregana)*, fairy bells *(Disporum smithii)*, violets *(Viola* spp.), yerba de selva *(Whipplea modesta)*, wild-ginger *(Asarum caudatum)*, Pacific starflower *(Trientalis latifolia)*, and members of the saxifrage family (Saxifragaceae).

One of the most aggressive tree competitors in coastal forests is the redwood *(Sequoia sempervirens)*, which allows few other trees to coexist with it, and beneath which relatively few shrub or herbaceous species are able to grow. This tree casts dense shade, has shallow roots that take up moisture and nutrients, produces a thick litter of branchlets on the ground that prevents seedlings from becoming established, and resists the fires and flooding that destroy competitors. The redwood is unusual in that it can regenerate new shoots after logging or severe fires, resulting in the so-called fairy rings of trees that are characteristic of forests that have been logged in the past.

Because of the great longevity of most coastal conifers, it is uncertain whether these forest communities are stable ones or whether their species composition changes over time. One rarely sees seedling redwoods except along road cuts or trails because seedlings of this conifer have high light requirements. It is likely, however, that prior to the development of the lumber industry along the Pacific Coast, frequent natural forest fires were an important restorative mechanism, removing the senescent older trees and producing the ecological conditions favored by their offspring.

Fortunately, visitors can see virgin stands of coastal conifers in a few places along the coast, notably the Olympic Peninsula and in California's redwood country.

Introduced Plants

Virtually all coastal plant communities have been invaded by plant species that are not native to North America. Some of these, such as European beachgrass *(Ammophila arenaria)* and Hottentot-fig *(Carpobrotus edulis)*, were introduced purposely to stabilize dunes but in the process have locally eradicated or reduced the native flora. Others such as brooms *(Genista* spp. and *Cytisus* spp.) and gorse *(Ulex europaea)*, foxglove *(Digitalis purpurea)*, and ox-eye daisy *(Leucanthemum vulgare)* were probably introduced as garden plants but escaped and have become very abundant in places. Brooms and gorse are considered as pests because they form very dense stands that crowd out native species. In addition, they are fire prone, and after fires their seeds germinate in abundance. Most coastal introductions probably arrived here accidentally as contaminants in seeds of crop or garden plants, as hitchhikers on clothing or in wool and in ballast or by unknown means. Included in this group are field mustard *(Brassica rapa)*, yellow parentucellia *(Parentucellia viscosa)*, sea-rocket *(Cakile maritima)*, and poison-hemlock *(Conium maculatum)*. Many of these so-called weeds probably do little if any harm to the native flora. Some of these species thrive in undisturbed habitats and are likely to persist as relative well-behaved residents of the Pacific Coast for the indefinite future.

Although coastal areas of the Pacific Coast states have been significantly altered by human activities, there are many places that are still botanically rich and varied. We hope that this book will enhance your enjoyment of coastal wildflowers and increase your appreciation of our great natural heritage.

Coastal counties of California, Oregon, and Washington

Leather-leaf fern

Although not wildflowers, the ferns and their allies are prominent plants of interest to many. Several kinds grow on rocky sea bluffs and in shaded and protected places down to the edge of the beach, especially northward. Several ferns are included in this book, such as **LEATHER-LEAF FERN** *(Polypodium scouleri),* which has a creeping, woody rootstock covered with loose scales. The leaves are thick, almost leathery, divided into blunt segments, and four to 16 inches long. Their undersurfaces bear naked, round, spore-producing clusters, or sori, crowded against the midribs. This fern is found on mossy logs, cliffs, and slopes from Santa Cruz Island, California, to British Columbia.

SILVERBACK FERN (*Pentagramma triangularis* subsp. *viscosa*) is a rather small fern that rises from a stout, ascending or short-creeping rootstock that is covered with brownish or blackish scales. The triangular fronds, or leaves, are sticky above and white powdery beneath. Borne on red brown stipes, or stems, they become quite curled up in the dry season. The sori are not conspicuous and are borne on the undersurface and along the veins. Silverback fern occurs on coastal slopes of San Diego and Orange Counties, California, and on several of the California islands.

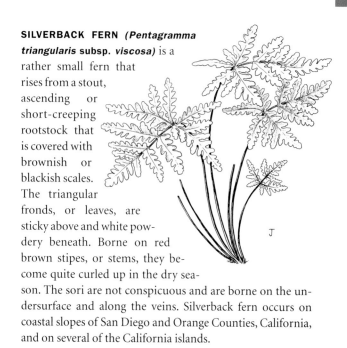

Formerly included in the genus *Cheilanthes*, **CALIFORNIA LACE FERN (*Aspidotis californica*)** is easily recognized by its finely divided triangular leaf blades, which are hairless and thickish in texture. They curl up tightly in dry weather, as do

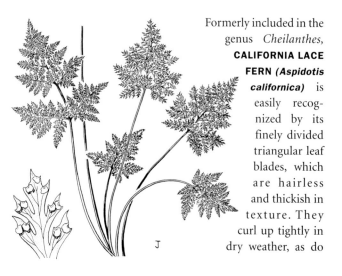

Western sword fern

those of the silverback fern *(Pentagramma triangularis* subsp. *viscosa)* described above. The fronds rise from a scaly, short-creeping rootstock, have wiry, dark brown stems, and reach a height of three to 12 inches. The sori are on the undersurface and protected by the revolute margins. California lace fern is found in rocky places from northern Baja California to Humboldt County, California.

WESTERN SWORD FERN *(Polystichum munitum)* is a species of cool, damp woods along the coast from Santa Cruz Island and Monterey County, California, northward to Alaska. It is a coarse, evergreen fern growing from woody, suberect, very scaly rootstocks. The many fronds are in heavy crowns or clumps two to four feet high and have stout

Bracken fern, or brake fern

stems that have conspicuous, chestnut brown scales. The leaves are pinnately divided, with the segments spreading to either side of the central midrib as in a feather or pinna. On the undersurface are many rounded sori, each covered by a flap of tissue.

BRACKEN FERN, or **BRAKE FERN,** *(Pteridium aquilinum* var. *pubescens)* has long-creeping, branched, hairy, underground rootstocks that send up stout fronds one to five feet tall. The stipes are straw colored, and the blades are three times divided and one to four feet long. The margin may be inrolled over the sori. This fern is widely distributed in California and northward to Alaska.

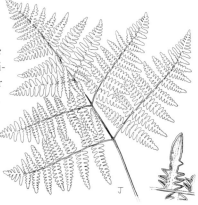

Distantly related to the ferns and also reproducing by spores instead of flowers is the horsetail, or scouring-rush. **GIANT HORSETAIL *(Equisetum telmateia* subsp. *braunii),*** has short-lived, jointed, unbranched stems that are either whitish or brownish. At the tip of each joint is a membranous sheath with 20 to 30 teeth. The summit of the stem bears a conelike

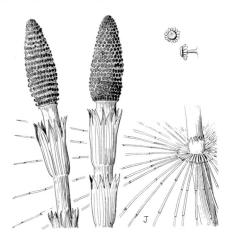

structure one to three inches long that produces the reproductive cells or spores. Later in the season, the sterile, green, much-branched stems appear. The stems and branches are hollow and jointed in this genus and solid only at the nodes, and the profuse branching is responsible for the common name of "horsetail." The plant grows in swampy and moist places from seepy banks on the beach to the interior and ranges northward to British Columbia.

COLUMBIA LILY (Lilium colum-bianum), in the lily family (Lil-iaceae), is not a beach plant but occurs in scrub and wooded places near the beach. It ranges from Humboldt County, California, to British Columbia, and blooms in June and July. The lower leaves are usually in whorls of five to nine, and the upper leaves are scattered. The perianth seg-ments (sepals and petals) are one-and-a-half to over two inches long, recurved, lemon yellow to golden to deep red with a yellow center, and usu-ally spotted maroon.

Columbia lily

YELLOW SKUNK-CABBAGE (Lysichiton americanum), like oth-ers of the arum family (Araceae), is characterized by having a greenish, central fleshy spike, or spadix, made up of many minute flowers and surrounded by a modified, in this case yellowish, leaf, or spathe. A spring bloomer, this plant has ill-smelling flowers and large, rather fleshy, leaves. It inhabits swampy or wet places along the coast from the Santa Cruz Mountains, California, northward to Alaska. Famil-iar members of this family are philodendrons (*Philo-*

Yellow skunk-cabbage

dendron spp.), calla-lily *(Zantedeschia aethiopica)*, and jack-in-the-pulpit *(Arisaena triphyllum)*.

GOLDEN-EYED-GRASS, or **YELLOW-EYED-GRASS,** *(Sisyrinchium californicum)*, like blue-eyed-grass *(S. bellum)*, is not a grass but a member of the iris family (Iridaceae). The plants are up to two feet tall and grow in very wet areas along the coast from California to British Columbia. The flowers are short lived and may remain closed on a foggy day, but on sunny days they enliven the landscape with their bright yellow, starlike flowers. Golden-eyed-grass is much less commonly encountered than is blue-eyed-grass. The name of the genus is Greek for "pig snout," supposedly because pigs dig the plants for their edible roots.

Golden-eyed-grass, or yellow-eyed-grass

HOTTENTOT-FIG *(Carpobrotus edulis)* is a member of the fig-marigold family (Aizoaceae), formerly known as the carpetweed family. An introduction from South Africa, it has been much planted along highways and banks to control erosion and has become naturalized on dunes and sandy places along the coast where it grows with sea-fig *(C. chilensis)*. The latter has rose magenta flowers, whereas the flowers of

Hottentot-fig

Hottentot-fig can be pink or yellow. Both have very fleshy three-sided leaves that are somewhat curved in Hottentot-fig and straight in sea-fig.

The buttercup family (Ranunculaceae) is especially important in temperate zones. True buttercups (*Ranunculus* spp.) have shining yellow petals, many stamens, and several central small pistils, each of which matures into a small, one-seeded dry fruit. In California, a common buttercup is **CALIFORNIA BUTTERCUP (*Ranunculus californicus*)**. A perennial with slender roots and with stems up to over two feet tall, it is a common plant in fields during spring months. It ranges from southern Oregon to northern Baja California and comes out to the coast in open places. A more strictly

California buttercup

coastal form with prostrate stems grows on coastal bluffs and hills of Santa Cruz and San Miguel Islands and along the coast from Monterey County, California, to Oregon. It is a small plant, and the stems attain a length of about 10 inches. Another similar species is western buttercup *(R. occidentalis)*, which usually has five to six petals one to two times as long as broad, whereas California buttercup has largely eight to 15 petals, mostly two to three times as long as broad.

Another coastal butter-cup is **MEADOW BUTTER-CUP *(Ranunculus acris)*,** a native of Europe and long since introduced into America and naturalized in moist places across the continent. It grows near the coast from Humboldt County, California, to Alaska. It is a more or less hairy perennial with several erect stems and is one-and-a-half to three feet tall and branched above. The five yellow petals are one-third to one-half inch long, and the more or less five-sided leaves are two to three inches wide and deeply cut. It has stout roots that persist from year to year.

CALIFORNIA MAHONIA, or **CALIFORNIA BARBERRY,** *(Berberis pinnata),* of the barberry family (Berberidaceae), is woody with prickly, somewhat hollylike leaves. The flowers are built on a plan of three: the six petals are in two series, and the nine sepals are in three series. The six stamens have flattened filaments. The fruit, a bluish berry about one-fourth inch long, has an acidic sap. The wood and inner bark are yellow and were once used for dye. This species is one of several members of the barberry family found on the Pacific Coast and grows in rocky, exposed places on Santa Cruz and Santa Rosa Islands and the mainland of California and Oregon.

California mahonia,
or California barberry

CALIFORNIA POPPY (Eschscholzia californica), of the poppy family (Papaveraceae), is widespread on the west coast, particularly in California, where it occupies many ecological

California poppy

niches, even on the bluffs and beaches along the sea. The flower is often yellow in plants near the coast rather than the familiar orange of inland plants. The typical form is heavy rooted, glaucous, and has smooth, broad multidissected leaves. It occurs from Santa Barbara to Mendocino Counties. In southern California, a prostrate form can also be seen from Monterey County south that has very gray, roughish, puberulent leaves that, under a lens, appear pitted when dry.

CREAM CUPS (Platystemon californicus) is a widely distributed hairy annual in the poppy family that grows in grassy areas, on rocky slopes, and on sand dunes from Baja California to Coos County, Oregon, and inland to Utah and Arizona.

Cream cups

As is true of all our native poppies, the sepals fall off when the flowers open. Cream cups commonly grow after an area has burned and can survive or even proliferate in fields that are lightly cultivated. The flower buds nod prior to opening, and the flowers have an odd odor. Despite the showy flowers, cream cups is wind pollinated rather than insect pollinated.

A shrub with an interesting distribution, namely deserts and coastal sea bluffs where conditions are somewhat saline, is **BLADDERPOD** *(Isomeris arborea),* of the caper family (Capparaceae). Everyone knows the caper of cookery, an Old World plant, but the American members of the family have mostly ill-smelling foliage, four-petaled flowers, and stalked, often inflated, seedpods. Bladderpod is a shrub up to several feet tall and has three-parted leaves and yellow flowers an inch or more across. Along the coast, this plant extends as far north as San Luis Obispo County, California.

Bladderpod

The mustard fam-
ily (Brassicaceae),
like the caper
family, has four-
petaled flowers
and peppery sap,
as seen in radish,
cabbage, and
other cultivated
plants. One of its
formerly com-
mon, but now less
frequently encoun-
tered, wild representa-
tives is **SAN FRANCISCO**
WALLFLOWER *(Erysimum franciscanum)*. The flowers are yel-
low to cream and over half-an-inch wide, and the seedpods
are an inch or more long, erect, and often tinged purple. A
short-lived, foot or more high perennial, this coastal species is
distributed from San Mateo County, California, to south-

San Francisco wallflower

western Oregon and is similar to other local species on sand dunes and bluffs fairly extensively distributed all along the Pacific Coast.

A more common wallflower is **COAST WALLFLOWER** *(Erysimum menziesii subsp. concinnum),* a biennial or short-lived perennial with generally dense rosettes of fleshy, subentire to sharply toothed leaves with branched hairs. The four cream to yellow petals can be over an inch long, and the erect seedpods are on pedicels about half-an-inch long. Generally found on headlands or cliffs, coast wallflower ranges from Pt. Reyes, California, to southern Oregon and flowers in winter and spring.

In the mustard family too is **FIELD MUSTARD**, or **TURNIP**, *(Brassica rapa)*, an erect annual one to three feet tall. Characterized by its clasping upper leaves and bright yellow flowers, it is an attractive plant, albeit an introduced weed from Europe. Being an annual, it is not excessively pernicious. It is widely distributed inland in orchards and fields but grows in sandy places and on bluffs along most of our coast, flowering in early spring in the south and into July farther north.

Field mustard, or turnip

All along our coast, especially on bluffs overlooking the sea, are succulents of the stonecrop family (Crassulaceae). Unlike the cacti, they are not spiny and have four or five sepals and petals and quite distinct pistils. The common name for the genus is live-forever, but *Dudleya*, its latin name, is also often used as its common name. There are a number of coastal species such as **SEA-LETTUCE** *(Dudleya caespitosa)*. Its rosetted leaves are two to eight inches long, and it bears clusters of bright yellow to red flowers that are about an inch wide. It is found from Monterey to Los Angeles Counties.

Sea-lettuce

Powdery dudleya

Closely resembling sea lettuce *(Dudleya caespitosa)* is **POW-DERY DUDLEYA** *(D. farinosa),* a species ranging from southern Oregon to Los Angeles County. It has pale yellow flowers, and the leaves are often gray glaucous and only one to two inches long. Other species occur farther south.

In the same family is **PACIFIC STONECROP** *(Sedum spathuli-folium).* It is a perennial with slender rootstocks and prominent rosettes of leaves reaching a length of about one inch. The erect or decumbent (prostrate below, erect above) flowering stems grow to about a foot high and bear branches of yellow to orange or white flowers half-an-inch or more across. Along the coast, this species grows from California to British Columbia and bears flowers from May to July.

In the rose family (Rosaceae) are the cinquefoils *(Potentilla* spp.), which often have leaves palmately divided (the segments spreading like the fingers from the palm of a hand).

Pacific stonecrop

Some species, however, have pinnately divided leaves (segments spreading to either side of the midrib like a feather), such as **PACIFIC SILVERWEED** *(Potentilla anserina* **subsp.** *pacifica).* This plant is a creeping perennial that has long runners or stolons (stems along the ground) and suberect leaves eight to 20 inches long with seven to 31 leaflets white woolly

Pacific silverweed

underneath. The bright yellow flowers are about one inch across and appear from April to August. It is found on sandy beaches and in salt marshes from southern California to Alaska.

The pea family (Fabaceae) has characteristic flowers consisting of a wide upper banner petal, two lower side wing petals, and a boat-shaped keel petal hidden between the wings. **COASTAL LOTUS (Lotus salsuginosus)** is a prostrate annual in this family and has smooth or slightly hairy stems to about one foot long and slightly fleshy leaves. The flowers are mostly yellow, becoming reddish with age, and about one-third inch long and have somewhat longer, straight seedpods. The species is found on the Channel Islands of California and in sandy places and on sea bluffs from Santa Clara County southward, spreading also into the interior.

Another coastal member of this genus is **RUSH LOTUS (Lotus junceus var. biolettii)**. This slender-stemmed perennial is much branched and somewhat woody at the base and has fine, appressed hairs and stems up to 20 inches long. The corolla is about one-fourth inch long and yellow

Marsh locoweed

tinged with red, and the seedpod is somewhat curved and beaked. It occurs on dry, coastal hills from Mendocino County, California, southward to San Luis Obispo County.

Locoweeds, or rattleweeds, (*Astragalus* spp.), are one of the large groups of the pea family in western North America, with well over 100 species in the Pacific states. In some areas they cause considerable poisoning of livestock. **MARSH LOCO-WEED** *(Astragalus pycnostachyus),* which has yellowish flowers, is a stout perennial more or less woolly with short twisted hairs. The erect stems grow to be almost three feet tall and bear large pinnately compound leaves. The flowers and seedpods are about half-an-inch long. Marsh locoweed once occurred in salt marshes or moist depressions behind dunes along much of the California coast; however, it is now limited to only a few places. (See "Whitish Flowers" for other *Astragalus* species.)

Another conspicuous coastal plant with pea-shaped flowers like coastal lotus *(Lotus salsuginosus)* described above is an in-

Furze, or gorse

troduction from Europe, namely, **FURZE,** or **GORSE (Ulex europaea)**. It is very densely branched with thick spiny branches, simple stiff spiny leaves, and showy yellow flowers. It is naturalized at spots along the coast from southern California to British Columbia but more abundantly in the north. In some regions, for instance, in Bandon, Oregon, this plant crowds out the native vegetation and forms an almost impenetrable mass. It flowers from February to July.

FRENCH BROOM (Genista monspessulana), also introduced from Europe, is a leguminous shrub from the Mediterranean region. French broom is now well established all along the California coast where the shrubs form large, dense stands in which native plants cannot compete. The hard-coated seeds are long lived, so removal of adult plants is usually followed by the appearance of vast numbers of seedlings. The shrub is attractive in flower, although the flowers and perhaps the foliage are said to be toxic. It has been reported that a shampoo to kill lice can be made from the dried tops of this plant if they are picked just before flowering.

Violets (*Viola* spp.) of the violet family (Violaceae) are a garden and wild favorite. The most coastal species is **EVERGREEN VIOLET *(Viola sempervirens),*** an almost hairless perennial from short, scaly rootstocks that produce creeping stems with scattered rounded leaves and lemon yellow flowers up to half-

Evergreen violet

an-inch long. The three lower petals are faintly purple veined. Growing mostly in shaded woods, it may occur down to the very edge of the beach and is distributed from central California to British Columbia.

Several species of the cactus family (Cactaceae) grow on coastal bluffs. One of these is **COASTAL PRICKLY-PEAR *(Opuntia littoralis).*** It is a large plant up to four or five feet high and has elongate joints a foot or so in length. The spines are whitish with red brown bases, and the flowers are yellow. It is distributed from Santa Barbara County, California, to Baja California. Other closely related plants occur inland.

Seaside fiddleneck

Fiddlenecks (*Amsinckia* spp.), members of the borage family (Boraginaceae) are annual, usually pungent-bristly herbs with yellow to orange corollas. A species found along the coast is **SEASIDE FIDDLENECK *(Amsinckia spectabilis)*,** an inhabitant of sandy places and the borders of salt marshes from Tillamook Bay, Oregon, to Baja California. It blooms from March to June, often growing in masses and usually more or less spreading or prostrate. The orange flowers are one-fourth to one-half inch long.

In the four-o'clock family (Nyctaginaceae) are found such plants as bougainvilleas (*Bougainvillea* spp.) and garden four-o'clock *(Mirabilis jalapa),* but the most showy of the native western members of this family are the sand-verbenas. Our beaches have three species, but there are others in the mountains, deserts, and on the plains of North America. **YELLOW SAND-VERBENA *(Abronia latifolia)*** is common on the coastal strand from Santa Barbara County, California, to British Columbia. The trumpet-shaped flowers have no petals, and the yellow petal-like parts are actually a modified calyx. The flow-

Yellow sand-verbena

ers occur in heads and have several sepal-like bracts (leaflike structures) below. Flowering is from May to October.

Another yellow-flowered plant is **SUN CUP *(Camissonia ovata)*,** of the evening-primrose family (Onagraceae). It is four petaled as in the mustard and caper families, but the ovary or seed-bearing part is inferior, that is, it is below instead of above the petals. In this species, the flower is at the summit of a long tube, and the ovary is hidden in the tuft of leaves at the base of the plant. Unlike many of the evening-primroses, sun cup is a day bloomer with bright yellow flowers highly reminiscent of the buttercup. It is found in open places along the coast from southern Oregon to San Luis Obispo County, California.

Sun cup

Beach-primrose

BEACH-PRIMROSE *(Camissonia cheiranthifolia),* like sun cup *(C. ovata),* is also a day bloomer. A perennial with more or less prostrate stems radiating from a central rosette of leaves, it is usually grayish hairy throughout and forms large mats in full maturity. The yellow petals may turn red in age and may be one-fourth to two-thirds of an inch long. Growing on the coastal strand, this plant occurs from Coos County, Oregon, southward. From Point Conception to northern Baja California, it is more woody and has larger flowers than those occurring north of Santa Barbara County.

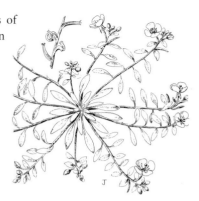

EVENING-PRIMROSE (*Oenothera elata*), for which the family is named, is a tall, rather weedy, biennial that typically lives over the first winter as a rosette of leaves and possesses a fleshy root. This is the large-flowered species found all through the southwestern states, opening in the late afternoon with flowers an inch or more in diameter. Two or three forms that differ in technical characteristics such as pubescence and length of sepal tips grow on moist beaches, seeps on the sea bluffs, and stream banks of much of the California coast.

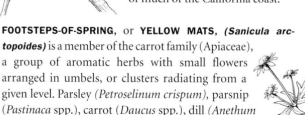

FOOTSTEPS-OF-SPRING, or **YELLOW MATS, (*Sanicula arctopoides*)** is a member of the carrot family (Apiaceae), a group of aromatic herbs with small flowers arranged in umbels, or clusters radiating from a given level. Parsley *(Petroselinum crispum),* parsnip *(Pastinaca* spp.), carrot *(Daucus* spp.), dill *(Anethum graveolens),* coriander *(Coriandrum sativum),* and anise

(Pimpenella anisum) are familiar members of this family. Footsteps-of-spring is a more or less prostrate perennial with three-parted leaves and small, yellow flowers that produce bristly fruits. This plant is found in sandy flats and on open hillsides mostly near the coast, from northern Oregon to central California.

Footsteps-of-spring, or yellow mats

MOCK-AZALEA (Menziesia ferruginea) belongs to the heath family (Ericaceae), together with plants such as rhododendrons (*Rhododendron* spp.), manzanitas (*Arctostaphylos* spp.), and blueberries (*Vaccinium* spp.). It is a deciduous shrub growing up to several feet tall and has glandular-hairy twigs, leaves one to two inches long, and yellow flowers tinged with red. The fruit is a dry capsule one-fourth inch long. Growing along the coast of Humboldt and Del Norte Counties, California, this species ranges to Alaska and Montana. It flowers in June and July.

A roadside weed near the coast from north-central California to Oregon is **YELLOW PARENTUCELLIA (Parentucellia viscosa),** of the important figwort family (Scrophulariaceae), which is known for plants such as snapdragon (*Antirrhinum* spp.), paintbrush (*Castilleja* spp.), monkeyflower (*Mimu-*

Yellow parentucellia

lus spp.), and penstemon (*Penstemon* spp.). This glandular annual has opposite, toothed leaves and terminal leafy spikes of yellow flowers with two-lipped corollas over half-an-inch long. An introduction from the Mediterranean region, this plant is abundant in disturbed places and blooms from April to June.

Another member of the figwort family is a native plant, **JOHNNY-NIP *(Castilleja ambigua)*,** which grows in low, saline places and on sea bluffs from British Columbia to Monterey County, California. The corolla ranges from one-half to one inch long and is yellow with purple markings. Found in Humboldt Bay is a form (*C. ambigua* var. *humbold-*

Johnny-nip

Common large monkeyflower

Bush monkeyflower, or sticky monkeyflower

tiensis), which has a purplish corolla and a yellow-tipped lower lip.

Monkeyflower, or *Mimulus,* is a large genus in the figwort family. One of the most widespread species is **COMMON LARGE MONKEYFLOWER (Mimulus guttatus),** a perennial, almost glabrous herb. The plants are stout, usually one to two-and-a-half feet tall, and the flowers are bright yellow with red spots, usually one-and-a-half to almost two inches long. Growing largely in wet places, especially seeps in coastal bluffs, this plant can be found along the coast from Santa Barbara County, California, to Washington.

Another species of *Mimulus* is **BUSH MONKEYFLOWER,** or **STICKY MONKEYFLOWER, (Mimulus aurantiacus).** Formerly in the genus *Diplacus,* along with other shrubby species, it is a profusely branched shrub, commonly two to four or five feet tall, glandular, and more or less viscid. The leaf veins on the upper surface are impressed, and the leaf edges are often turned under. The flowers are deep orange to yellow orange and one-and-a-half to almost two inches long. It grows in rocky places often on the immediate coast, from western Oregon to south-central California, and blooms from March to August.

CALIFORNIA BEDSTRAW (Galium californicum), of the madder family (Rubiaceae), ranges from Humboldt to Los Angeles Counties, California. It has slender, creeping rootstocks and tufted, slender stems to about one foot long that have recurved prickles and whorls of four leaves one-fourth to one-half inch long.

The staminate (male) flowers occur largely in groups of two or three; the pistillate (female) flowers are solitary. The corolla is yellowish and very small. The fruit is fleshy and smooth or hairy, becomes white when ripe, and blackens as it dries.

Twinberry

TWINBERRY *(Lonicera involucrata* var. *ledebourii)* is in the honeysuckle family (Caprifoliaceae), a woody family with united petals often forming two-lipped corollas. It is an upright shrub, and this coastal variety is three to 10 or more feet tall. The flowers are in pairs arising from the leaf axils (angle formed where the leaf stem joins the main stem) and have a yellowish corolla often tinged red, half-an-inch or longer. Two broad fused bracts, which may become purplish or reddish, almost enclose the black fruit. Flowering from March to April, it ranges from Santa Barbara County, California, to Alaska.

The sunflower family (Asteraceae) is noteworthy because of its great size, with about 21,000 species worldwide. In this family, the many florets are produced in a head with an involucre below formed by a series of leaflike bracts. This head tends to resemble the solitary flower of other families. The florets may all be alike and strap shaped (ray flowers) as in common dandelion *(Taraxacum officinale)*, all tubular (disk flowers) as in brass buttons *(Cotula coronopifolia)*, or of both kinds as in **RACEMOSE GOLDENBUSH,** or **RACEMOSE PYRROCOMA, (Pyrrocoma racemosa),** pictured here. This perennial has a short taproot and a tuft of basal leaves and bears several stems up to three feet long. The several to many heads

consist of outer ray flowers, which are elongate and petal-like and to about half-an-inch long, and many central disk flowers. The species occurs in coastal salt marshes and adjacent areas from Oregon to central California.

COASTAL GOLDENBUSH, or **COASTAL ISOCOMA, *(Isocoma menziesii* var. *vernonioides)*** ranges from San Francisco southward. It is a shrub up to about three feet high, somewhat resinous, and very leafy. The heads have no ray flowers, only the yellow tubular disk flowers that are characteristic of the centers of the heads of so many members of the sunflower family. One of these disk flowers is shown on the right side of the drawing, illustrating the one-seeded ovary at the base; the hairy modified sepals, or pappus; the tubular corolla with five lobes representing the petals; and the two-lobed stigma at the summit.

BUSH-SUNFLOWER (Encelia californica) is a low, rounded sub-shrub with green leaves one to two-and-a-half inches long and showy sunflower-like heads two to three inches across.

The outer petal-like ray flowers are yellow, and the central tubular disk flowers are purplish brown. It inhabits coastal bluffs and inland canyons from Santa Barbara County, California, to Baja California, blooming from February to June.

Bush-sunflower

Likewise in the sunflower family is the genus *Coreopsis*, of which several species are cultivated. Along the California coast are two perennials of some size and with stout stems and fleshy leaves divided into linear segments. The large, yellow flower heads are two to three inches across. The first of these is **SEA-DAHLIA (Coreopsis maritima)**, which grows on coastal bluffs and dunes of San Diego County and of northern Baja

Sea-dahlia

California and the adjacent islands. An herbaceous perennial, this plant has many stems one to almost three feet long, leaves two to 10 inches long, and few heads on stout naked stems six to 20 inches long. It blooms from March to May.

GIANT COREOPSIS (*Coreopsis gigantea*) is also a spring bloomer. The plant is shrubby, usually with one trunk three to several feet high, simple or branched, and bears terminal tufts of leaves two to 10 inches long and clusters of yellow-flowered heads. In the dry season, flowers and leaves are shed and all that remains is a dead-looking stick. This plant ranges from San Luis Obispo to Los Angeles Counties and is also found on most of the Channel Islands.

Giant coreopsis

COMMON WOOLLY-SUNFLOWER *(Eriophyllum lanatum)* is a most-variable perennial with a somewhat woody base and stems varying in height. The leaves are usually toothed or divided and one-half to three inches long. The flower heads of bright yellow petal-like ray flowers are solitary or in open clusters with hemispheric involucres one-fourth to one-half inch high. Because there are several named varieties, more than one of which occurs along the coast, the species in its various forms may range from British Columbia to central California.

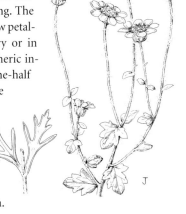

SEASIDE WOOLLY-SUNFLOWER, or **LIZARD TAIL,** *(Eriophyllum staechadifolium)* is shrubby, much branched, one to five feet high, and woolly, especially when young. The rather narrow

Common woolly-sunflower

Seaside woolly-sunflower, or lizard tail

leaves may be entire, few lobed, or pinnatifid (deeply divided, almost to the midrib) and are permanently white woolly beneath. The numerous heads are in rather dense clusters, the involucres are about one-fourth inch high, and the outer yellow ray flowers are one-sixth inch long. Growing along the coast on beaches and bluffs from Coos County, Oregon, to Santa Barbara County, California, it blooms from April to September.

A quite different *Eriophyllum* is **MANY-STEMMED ERIOPHYL-LUM *(Eriophyllum multicaule)*.** It is a loosely woolly annual, one to six inches high, and diffusely branched. The leaves are almost wedge shaped, toothed or lobed at the apex, and about a third of an inch long. The boat-shaped leaflike bracts below the flower head loosely enclose the outer achenes. The yellow ray flowers are very short. This plant grows in old, sandy fields and on dunes from Monterey to San Diego Counties and flowers in spring.

AMBLYOPAPPUS *(Amblyopappus pusillus),* likewise of the sunflower family, has very small heads without ray flowers, all the corollas being minute, tubular disk flowers. The plant is a yellow green, balsamic, slender annual three to 14 inches high and bears slender, entire or somewhat divided leaves. It occurs on beaches, old dunes, and sea bluffs from San Luis Obispo County, California, to Baja California and on the Channel Islands. It flowers from March to June.

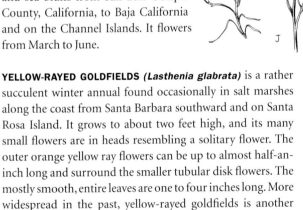

YELLOW-RAYED GOLDFIELDS *(Lasthenia glabrata)* is a rather succulent winter annual found occasionally in salt marshes along the coast from Santa Barbara southward and on Santa Rosa Island. It grows to about two feet high, and its many small flowers are in heads resembling a solitary flower. The outer orange yellow ray flowers can be up to almost half-an-inch long and surround the smaller tubular disk flowers. The mostly smooth, entire leaves are one to four inches long. More widespread in the past, yellow-rayed goldfields is another example of how recent human activity is depleting many

Yellow-rayed goldfields

of our native wildflower populations.

Much more frequently encountered, both on the coast and inland, is **COMMON GOLDFIELDS (Lasthenia californica),** a small winter annual with pairs of narrow leaves somewhat sheathing at the base. The yellow heads have conspicuous rounded ray flowers one-fourth to almost one-half inch long and several disk flowers in the center. Below the head, several leaflike bracts, separate and overlapping, form a hemispheric involucre. This species is colonial, forming great masses of yellow along practically the length of California and southwest Oregon. Flowers come from March to May.

Common goldfields

Also found infrequently near the coast is **PERENNIAL GOLD-FIELDS (Lasthenia macrantha),** a species that lasts more than one year and is more or less hairy with ascending hairs. The leaves are one to almost eight inches long, narrow, and paired as in yellow-rayed goldfields (*L. glabrata),* but the ray flowers  are slightly longer. This plant occurs in somewhat different forms in grassy places and on dunes from Curry County, Oregon, to San Luis Obispo County, California. Flowering is mostly from March to August.

One of the first native annuals to flower in spring is **BLENNOSPERMA (Blennosperma nanum),** a member of the sunflower family known to many but having no common name other than its genus name. The variety *B. nanum* var.

Blennosperma

nanum grows to 10 inches tall in meadows along the coast from Marin to San Diego Counties, California, although it is rare in southern California. *B. nanum* var. *robustum,* as its name implies, is slightly taller and has larger flower heads. It is restricted to Point Reyes in Marin County. The generic name is Greek for "sticky seeds"; when the dry seeds become wet they soon are enveloped by a mucilaginous covering whose function is unknown. A second species, Baker's blennosperma *(B. bakeri)*, occurs only in a small area just north of San Francisco Bay and can be distinguished by its red rather than yellow stigmas.

Yellow pincushion, or pincushion flower

Another sunflower family plant that varies in different parts of its range is **YELLOW PINCUSHION,** or **PINCUSHION FLOWER, *(Chaenactis glabriuscula),*** a winter annual four to 16 inches high. Its many small, yellow flowers are arranged in a compact head. All of the flowers are tubular disk flowers, the outer somewhat enlarged. The species varies from almost smooth to quite woolly and from large headed to smaller, and its leaves are much divided to less divided. This flower occurs in the interior and coastal regions of much of California and comes down to the beach in more than one form and at various places from the San Francisco Bay Area to San Diego.

Coast sneeze-
weed

COAST SNEEZEWEED *(Helenium bolanderi)* forms clumps one to two feet high with several stout stems from a thick root. It is more or less woolly, especially about the long-stemmed, solitary heads that have hemispheric centers of tubular disk flowers and outer, yellow petal-like ray flowers to about an inch long. It is frequent in moist meadows and coastal swamps and on bluffs from Mendocino County, California, to Coos County, Oregon.

In salt marshes and wet places on the beach, and scattered from northern Baja California to the Puget Sound region and Vancouver Island, is **JAUMEA** *(Jaumea carnosa),* a fleshy perennial with creeping branched rhizomes (underground stems) and numerous, mostly simple, more or less prostrate or ascending stems. The heads are generally solitary and have narrow, often inconspicuous ray flowers and central, yellow

Jaumea

disk flowers. The leaflike bracts below the flower heads are quite fleshy and characteristic of this species.

Another quite attractive member of the sunflower family is **CANYON-SUNFLOWER** *(Venegasia carpesioides),* a leafy, more or less hairless perennial that has many stems and thin, bright green leaves two to six inches long. The leaflike bracts below

Canyon-sunflower

the flower head are loosely arranged, and the outer bracts are spreading. The bright yellow petal-like ray flowers are almost an inch long and obscurely toothed at the blunt tip. This species is largely a shade plant of rocky and steep places along the coast from Monterey County, California, to northern Baja California.

Gumplant, or grindelia, is a perennial with a gummy or resinous exudation, especially from the heads. **PACIFIC GRINDELIA**, or **PACIFIC GUMPLANT**, *(Grindelia stricta* var. *platyphylla)* is a plant of coastal marshes and seaside bluffs and ranges from Coos County, Oregon, to Monterey County, California. It is a more or less trailing perennial up to three feet across and has yellowish or whitish stems and finely toothed, gland-dotted leaves. The heads are from one to almost two inches across, and the stiff, leaflike bracts below are recurved at the tips.

Quite a large western group of the sunflower family is known as the tarweeds because of their glandular or sticky and

Pacific grindelia, or Pacific gumplant

Woodland madia, or woodland tarweed

heavy-scented herbage. One found along the coast is **WOOD-LAND MADIA,** or **WOODLAND TARWEED,** *(Madia madioides)*. It is a slender-stemmed perennial one to two feet tall and forms a well-developed basal rosette of coarsely hairy, slightly toothed leaves two to four inches long. The heads are few, yellow, and showy. This species is found largely in coniferous woods along the coast from Monterey County, California, to Vancouver Island. It blooms from July to September.

Goldenrod is a group of perennial herbs with leafy, usually simple stems and alternate leaves. **WEST-ERN GOLDENROD *(Euthamia occidentalis)*** is stout, three to six feet tall, and glabrous and bears lance-linear leaves two to four inches long. The flowers are in small heads with 15 to 25 short, outer ray flowers and six to 15 tubular inner disk flowers. Not primarily a shore inhabitant, this gold-

enrod does reach the coast in moist places and ranges from British Columbia to Baja California and Texas. It blooms from July to November.

A typical member of the sunflower family is **DUNE TANSY** *(Tanacetum camphoratum)*. A perennial from stout, underground rhizomes, it becomes one to two feet high and has dissected leaves to about eight inches long. The heads have numerous tubular disk flowers and very short but more or less evident ray flowers. The plant is more or less hairy but not whitish. It is found on the coastal strand from Humboldt County, California, to British Columbia. A unique form of dune tansy that is white woolly, especially on younger parts, and whose ray flowers are not at all evident is found around San Francisco Bay.

Dune tansy

Brass buttons

BRASS BUTTONS *(Cotula coronopifolia)* is a common annual plant found in saline or freshwater marshes along the entire California coast (and inland as well). It often grows in large, dense colonies and produces a showy display when its golden yellow flower heads, consisting of only disk flowers, are produced. The leaves emit a distinctive, aromatic fragrance when bruised. This member of the sunflower family was introduced from southern Africa, probably unintentionally, but has taken a firm hold in California, as well as in other regions that have a similar climate, such as parts of Australia. The name "brass buttons" alludes to the yellow heads and probably was coined after its arrival in California; in southern Africa it goes by the names "goosegrass" or "duckling-plant."

One of the truly large genera of flowering plants is *Senecio*, often called groundsel, butterweed, or ragwort. This genus has well over 1,000 species; some are trees, others are vines and shrubs, but most are herbs. On our coast, growing on dunes and back beaches of San Luis Obispo and Santa Barbara Counties, California, is **BLOCHMAN'S BUTTERWEED,** or

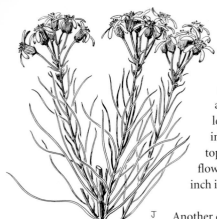

BLOCHMAN'S SENE-CIO *(Senecio bloch-maniae)*. It is a subshrub that grows up to about three feet high and has linear-filiform leaves one to three inches long and flat-topped groups of yellow-flowered heads about one inch in diameter.

Another quite different member of the *Senecio* genus is **SEACOAST BUTTERWEED *(Senecio bolanderi* var. *bolanderi),*** a perennial herb of the immediate coast that has underground rootstocks. Its slender stems are one to two feet tall and produce rounded or somewhat heart-shaped leaves at the base, whereas the leaves on the stem are more lobed. The flower heads are about one-third inch high and one inch across. It is found on the coastal strand and neighboring bluffs from Washington and southwestern Oregon to Mendocino County, California. Another coastal species is California butterweed (*S. californicus*), a low annual with leaves clasping the stem, occurring from Monterey County, California, southward.

One of the most specialized groups in the sunflower family has all the florets modified into elongate, strap-shaped ray flowers. Among these is **COAST MICROSERIS (Microseris bigelovii),** a stemless annual of coastal bluffs and flats from Santa Barbara County, California, to British Columbia. The leaves are two to 10 inches long and entire or with tooth-like, lateral lobes. The heads are solitary at the ends of long stems, and each consists of 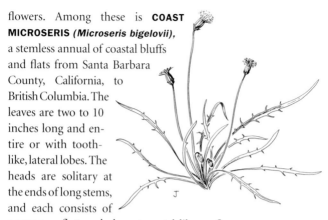 numerous, flattened, elongate, petal-like ray flowers.

Related to microseris (*Microseris* spp.) in having all ray flowers are agoseris, or mountain dandelions (*Agoseris* spp.). Commonly found in the pine belt of western mountains, a species along our beaches and their environs is **SEASIDE AGOSERIS (Agoseris apargi-oides),** which occurs from southwestern Washington to Santa Barbara County, California. It is a perennial and thinly hairy to woolly or almost 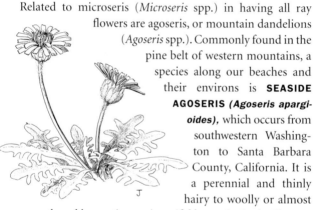 smooth and has entire to pinnatifid leaves (as in the illustration) and dandelion-like heads an inch or more across.

Easily confused with seaside agoseris is **CAT'S EAR,** or **FALSE DANDELION, (Hypochaeris radicata),** a showy, dandelion-like plant. A perennial with lobed, rough-haired leaves two to six inches long and stems one to three feet high, it has bright yellow heads an inch or more across made up of strap-shaped

Seaside agoseris

Cat's ear, or
false dandelion

ray flowers. It can be distinguished from the similar-looking
seaside agoseris by its lack of hairs on the phyllaries, its often
branched stems, and the scale at the base of each ray flower.
Introduced from Europe, it blooms much of summer but is

Corn chrysanthemum

very conspicuous in June in open places in woods, on sea bluffs, and along trails and roadsides in much of California and northward, as well as across the continent.

The chrysanthemum group of the sunflower family was formerly a large genus of the northern hemisphere but is now divided into three separate genera *(Argyranthemum, Chrysanthemum,* and *Leucanthemum)*. Plants in this group have two to four series of overlapping, leaflike bracts below the flower head, and many species are quite aromatic. **CORN CHRYSANTHEMUM** *(Chrysanthemum segetum)* is a yellow-flowered annual naturalized along the coast and in fields of coastal central and northern California. This chrysanthemum is one to two feet high and much branched; its leaves are cut or deeply divided, usually with a clasping base; and its heads are one to two inches across. It is a native of the Mediterranean region.

COAST LILY *(Lilium maritimum)* is a member of the lily family (Liliaceae), to which many of our common garden plants belong. This lily has an elongate bulb and stems one to four or more feet high. The leaves are usually scattered, not whorled, dark green, and one to five inches long. The flowers are horizontal, bell shaped, dark red, maroon spotted, and about one-and-a-half inches long. Found in sandy soil or on raised hummocks in bogs or in brush and woods, it ranges from Marin to Mendocino Counties, California. The more common leopard lily *(L. pardalinum)* also occurs near the coast in moist places and along stream banks. Flowers are pale orange to red at the tips, with maroon spots; central leaves are whorled.

Coast lily

SLINK POD, FETID ADDER'S TONGUE, or **BROWNIES** (as it is known in Humboldt County), *(Scoliopus bigelovii)* is also in the lily family. Although two of these common names have an ominous ring, slink pod is one of the most interesting plants found in the redwood forests from Humboldt to Santa Cruz Counties, California. In very early spring (as early as February), when the pair of blotched, pale green leaves emerge from the ground, the very small but intricately structured green and purple flowers are produced on short stems. These flowers have a foul odor that is apparently attractive to flies, who are

Slink pod, fetid adder's tongue, or brownies

the chief pollinators of this forest-floor denizen. After pollination, the flower stem elongates, pushing the ripening fruit along the ground away from the mother plant (hence the name "slink pod"). Its flowers are a favorite food of banana slugs, which may eat all the flowers from large numbers of adjacent plants. The smaller Oregon fetid adder's tongue *(S. hallii)* grows in forests of coastal mountains from southern to central Oregon.

Checker-lily, or mission bells

The curious **CHECKER-LILY,** or **MISSION BELLS,** *(Fritillaria affinis* **var.** *affinis)* is a member of the lily family that is common along the Pacific Coast between the San Francisco Bay Area and British Columbia and flowers as early as March. Its odd purplish, mottled flowers are often overlooked by unobservant humans, but deer seek out the flowers as a delicacy and may remove an entire season's flowers during a single mealtime visit. Although the flowers are nearly without fragrance, they are visited by flies, who may be attracted to them because of their dunglike or fleshlike coloration, and during their visits these flies pollinate the flowers. The seed capsules are odd, winged, papery structures.

COASTAL ONION (*Allium dichlamydeum*) is a handsome, small, native onion that grows on clay or rocks of dry, sea cliffs in central and northern California. Its odor and taste make this plant unmistakably an onion. Not surprisingly, the distinctive odor of onions is due to sulfur-containing compounds. Like onions that are edible to humans, this one has a bulb, but it is less than half-an-inch long. The flower stalks are up to a foot tall, but the one to three flat, shorter leaves lie on the ground.

Coastal onion

Stream orchid

STREAM ORCHID (*Epipactis gigantea*), of the orchid family (Orchidaceae), has the characteristic inferior twisted ovary of that family. It has a creeping rootstock with fibrous roots, simple leafy stems to almost three feet tall, and flowers with greenish, deeply concave sepals and purplish or reddish petals. The lip is strongly veined and marked with purple or red. Growing along moist stream banks at low elevations, it is sometimes found right on the shore; its full range is from Baja California to British Columbia, South Dakota, and Texas.

WILD-GINGER (*Asarum caudatum*), of the pipevine family (Aristolochiaceae), to which the mostly tropical dutchman's pipe (*Aristolochia* spp.) belongs, is low growing and has slender, spicy-smelling rootstocks and foliage. The leaves are evergreen and one to four inches long, and the brownish

Wild-ginger

purple flowers appear hidden near the base of the plant from May to July. The plant grows in deep shade and may occur in woods right down to the edge of the beach. It ranges from central California to British Columbia.

AUSTRALIAN SALTBUSH (*Atriplex semibaccata*) belongs to the goosefoot family (Chenopodiaceae), which is known for its small, inconspicuous flowers that lack petals. Plants such as beet (*Beta vulgaris*), spinach (*Spinacia oleracea*), and Russian-thistle (*Salsola tragus*) are in this family. Different species of saltbushes (*Atriplex* spp.) are largely identified by the shapes of their seed-encasing bracts (see the upper right-hand corner

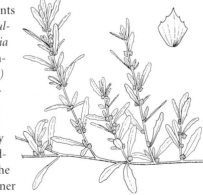

of the illustration). Australian saltbush is a prostrate perennial with scurfy, grayish hairs and reddish, fleshy, fruiting bracts. It has become established in saline places in the interior and along the coast from central California southward.

Formerly of the fumitory family (Fumariaceae), but now included in the poppy family (Papaveraceae), is the genus *Dicentra,* in which the outer pair of the four petals is saclike or spurred at the base. **BLEEDING HEART** *(Dicentra formosa)* is generally a plant of shady, wooded places but is sometimes found on the immediate coast. It ranges from central California to western British Columbia and flowers from March to July. When ripe, the fleshy capsules explode, scattering the seeds widely.

Bleeding heart

PACIFIC CAMPION, or **PACIFIC CATCHFLY, *(Silene scouleri* subsp. *grandis)*** is a member of the pink family (Caryophyllaceae). It is a perennial that has one to few stems rising from a heavy crown and leaves one to six inches long, and it can be up to two feet tall. The 10-nerved calyx is glandular and about half-an-inch long, and the divided petals are rose to greenish white. This plant grows on bluffs along the coast from San Mateo County, California, to British Columbia.

The purslane family (Portulacaceae) is usually fleshy and has two sepals. **SEASIDE CALANDRINIA *(Calandrinia maritima)*** is a glaucous annual with more or less spreading stems up to 10 inches long and leaves that are largely near the base of the plant. The sepals are rounded and dark veined, and the red petals are about one-fourth inch long. It grows in sandy places and on sea bluffs from Santa Barbara County, California, to northern Baja California and bears flowers from March to May.

SEA-FIG *(Carpobrotus chilensis)*
belongs to the fig-marigold family
(Aizoaceae) and has trailing stems
bearing opposite three-sided leaves.
The rose magenta flowers are one to
two inches broad, whereas in Hotten-
tot fig *(C. edulis)*, they are yellow, dry
pink, and are three to four inches
across. The sea-fig ranges from Oregon to Baja California. Both species
are introduced and have commonly
been used along road cuts and banks to
prevent erosion.

The mustard family (Brassicaceae), with its four sepals and
petals and superior ovary, is common and has several repre-
sentatives along the coast. A fleshy, branched, and glabrous
species is **SEA-ROCKET** *(Cakile maritima)*. Its seedpod is fleshy
and transversely two jointed. This plant, introduced from Europe, is found on beach sand from Monterey County to Mendocino County, California. Its leaves are pinnatifid (deeply divided, almost to the midrib), and the petals are almost
half-an-inch long. Another species is California sea-rocket *(C.
edentula),* now rarely encountered but once quite common.

Sea-rocket

Coast
rock-cress

The leaves are merely wavy toothed, and the petals are one-fourth inch long. This plant's common name is perhaps misleading because it was introduced here from the east coast. It is found on the west coast from San Diego to British Columbia.

COAST ROCK-CRESS (*Arabis blepharophylla*), another member of the mustard family, is a low perennial with one to a few simple stems, and its coarsely hairy lower leaves are in rosettes. The sepals are oblong, purplish, and about one-third inch long, whereas the rose purple petals measure half-an-inch or longer. The erect, smooth seedpods are an inch long or more. This pretty little species is found in rocky places along the coast from Santa Cruz County to Sonoma County, California, and flowers from February to April.

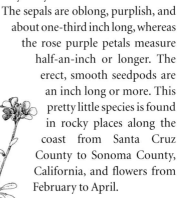

RED SAND-VERBENA (*Abronia maritima*), of the four o'clock family (Nyctaginaceae), is a fleshy, much-branched, prostrate, and sticky-haired plant much like yellow sand verbena (*A. latifolia*). The small flowers, however, are dark crimson to

Red sand-verbena

red purple and about one-sixth inch wide, whereas in the other species they are yellow and as much as one-third inch across. Red sand-verbena is found on the lower coastal strand from Baja California to San Luis Obispo County, California, and blooms from February to October.

THRIFT, or **SEA-PINK, *(Armeria maritima* subsp. *californica)*** belongs to the leadwort family (Plumbaginaceae). It is a tufted perennial with persistent, basal, linear leaves and naked

Thrift, or sea-pink

stems three to 15 or more inches high that bear heads of rose pink funnel-shaped flowers about one-third inch in length. It grows on coastal bluffs and in sandy places from San Luis Obispo County, California, to British Columbia. The flowers appear from April to August.

The morning glory family (Convolvulaceae), composed mostly of trailing or climbing plants, is a large family of warmer regions. The true morning glory has a number of species near the coast, such as **BEACH MORNING GLORY (Calystegia soldanella),** a fleshy prostrate perennial from rootstocks deep seated in the beach sands. The kidney-shaped, shiny, fleshy leaves are one to two inches wide, and the rose to purplish corolla is one-and-a-half to two-and-a-half inches long. The species is common on the coastal strand from San Diego to Washington and occurs also in South America and Europe, flowering in our area from April to August.

Beach morning glory

In the mint family (Lamiaceae), hedge-nettles, or members of the genus *Stachys,* are usually found in the West in damp places. A coastal species, especially in seeps and similar places on bluffs and in canyons, is **CALIFORNIA HEDGE-NETTLE (Stachys bullata).** A perennial with slender stems that are simple or branched and one to three feet high, it has stiff hairs bent downward on the stem angles. The leaves are one to six inches long, and the purple flowers are in whorls of six and one-half to almost one inch long. This plant and closely related species extend along much of the Pacific shore.

California hedge-nettle

In the figwort family (Scrophulariaceae) is **CALIFORNIA BEE PLANT,** or **CALIFORNIA FIGWORT,** *(Scrophularia californica).* It is a coarse perennial and three to five feet tall and has large, more or less triangular leaves and numerous small flowers in

California bee plant, or California figwort

large, terminal panicles. The corolla is red brown to maroon and about half-an-inch long. The species is found near the coast, often in brushy and damp places from Los Angeles County, California, to British Columbia. It blooms from February to July. Other forms occur away from the coast.

Also in the figwort family is **SALT MARSH BIRD'S-BEAK (Cordylanthus maritimus),** a branched annual, often with the stems more or less prostrate or nearly so. It is hairy, and some of the hairs are gland tipped. The bright green leaves and bracts are glaucous and up to an inch long. The tubular calyx is half-an-inch to almost one inch long and has terminal short sharp teeth, and the more or less purplish corolla is also tubular. The plant occurs in a number of the remnant salt marshes found along the coast from southern Oregon to northern Baja California and blooms from May to October.

Salt marsh bird's-beak

A species introduced from Europe and now well established in more or less shaded places near the coast from Santa Barbara County, California, to British Columbia is **FOXGLOVE (Digitalis purpurea)**. Also a member of the figwort family, it is a stout biennial two to six feet high and has large lower leaves

Foxglove

and terminal clusters of showy purple to whitish flowers. The corolla is nodding, somewhat inflated below, and about two inches long. Flowering is from May to September.

There are many western species of paintbrush (*Castilleja* spp.), and several are found along the immediate coast. Paintbrushes also belong to the figwort family, and they have long, narrow corollas often with very short, lower saclike lips. Most of the color is in the bracts that are situated below each flower. **MONTEREY INDIAN PAINTBRUSH** *(Castilleja latifolia)* has leaves that are less than three times as long as wide and blunt and sessile. The corolla is about an inch long. This species is found in sandy places along the coast of northern California, whereas closely related ones appear farther south and as far north as Washington.

Monterey Indian paintbrush

Woolly Indian paintbrush

Another somewhat woody paintbrush is **WOOLLY INDIAN PAINTBRUSH *(Castilleja foliolosa)***. It is bushy, white woolly throughout, and one or two feet tall and has narrow leaves, the uppermost having one or two pairs of lobes. The corolla is scarcely an inch long. This species is found in dry, rocky places in the Coast Ranges from Humboldt County, California, to northern Baja California and, to some extent, even farther inland. A plant somewhat similar in its woolliness is white-felted Indian paintbrush *(C. lanata* subsp. *hololeuca),* but it is only found in the northern Channel Islands. The upper corolla lip is yellowish with pale, thin edges instead of greenish with reddish edges as in woolly Indian paintbrush.

The saxifrage family (Saxifragaceae) has a tube at the base of the flower and four or five sepals and petals. In this family

Pig-a-back plant

are some common plants such as alumroot *(Heuchera* spp.), woodland star *(Lithophragma* spp.), and California saxifrage *(Saxifraga californica).* **PIG-A-BACK PLANT *(Tolmiea menziesii)*** is a perennial herb with scaly rootstocks and chiefly basal leaves. The flowers are in elongate clusters with five unequal sepals (three long and two short), a purplish tube, and five brownish, elongate threadlike petals. It reaches the coast in moist cool places from northern California to Alaska and flowers in May and June.

Related to pig-a-back plant is **FRINGE CUPS *(Tellima grandiflora),*** another plant with horizontal rootstocks and both basal and stem leaves. The plant is quite hairy and one to two-and-a-half feet high. The bell-shaped flower is about one-sixth inch long, with one-fourth-inch-long petals that are whitish at first and later red. This species is not primarily a

Fringe cups

shore plant but reaches the coast in wooded places and ranges from central California to Alaska. It flowers from April to June.

In the gooseberry family (Grossulariaceae) is **FUCHSIA-FLOWERED GOOSEBERRY *(Ribes speciosum)*,** a red-flowered spiny shrub. It is remarkable among currants and gooseberries because it has only four sepals and petals instead of the usual five. The bush is more or less evergreen, three to six feet tall, and very spiny and has glossy, deep green leaves. The bright red hanging flowers are striking and appear from January to May. It ranges on coastal bluffs and in adjacent canyons from Santa Clara County, California, to northern Baja California.

NOOTKA ROSE *(Rosa nutkana* var. *nutkana)*, of the rose family (Rosaceae), is another plant that is not primarily a shore plant but reaches the coast in northern wooded areas. It is a stout-stemmed plant, mostly armed with straightish heavy

Fuchsia-flowered gooseberry

prickles, and grows to a height of three to six feet. The fragrant pink flowers are three or more inches wide, and the rounded fruits are over half-an-inch in diameter. As its name indicates, it grows in southern Alaska and southward along the coast into northern California, where it can be found in damp soils of forest openings. In southern California, the common **CALIFORNIA WILD ROSE** *(Rosa californica)* often reaches the coast.

Nootka rose

This rose can be distinguished by its entire, not toothed, sepals and its recurved prickles. Specimens that combine characteristics of more than one species are occasionally found, suggesting that our native rose species hybridize with each other.

SALMONBERRY (*Rubus spectabilis*) is a blackberry relative that occurs from the Santa Cruz Mountains to Alaska. The common name may allude to its fruits, which are often the same orange color as the eggs of the salmon that spawn, or once spawned, in the coastal streams and rivers along which this shrub is so common; or it may allude to the fact that Native Americans along the northwest coast once ate its berries mixed with salmon eggs. Although the berries are edible to humans, to some palates they are at best insipid and at worst

Salmonberry

Coast clover, or cow clover

distasteful. In early spring, the beautiful reddish purple flowers produced on the bare twigs of this member of the rose family are a welcome sign that winter is over.

The clovers belong to the pea family (Fabaceae) and are characterized not only by their pea-shaped flowers and compound leaves, usually of three leaflets, but also by their short, generally one- to two-seeded pods. A clover that can be found on the immediate coast is **COAST CLOVER,** or **COW CLOVER, (Trifolium wormskioldii),** named for its Danish discoverer. It is a perennial from creeping rootstocks and has branching, rather coarse stems. The flowers are about half-an-inch long and whitish to purplish red. This clover occurs in wet places and is quite common in seeps on rocky bluffs from central California to British Columbia.

Another member of the pea family is **ROUND-LEAVED PSORALEA (Hoita orbicularis),** which is found in moist places in much of California, including along the shore. It has prostrate stems with long leaf stems up to 20 inches high, leaflets one to three inches

Round-leaved psoralea

long, and flower stalks one to two feet tall. The flowers are reddish purple, over half-an-inch long, and appear in early summer. The heavy-scented foliage is gland dotted and sometimes hairy.

SAND PEA *(Lathyrus japonicus)* is a perennial beach pea that has well-developed tendrils like most species of *Lathyrus*. The leaves are green and more or less fleshy. The flowers are two to eight in number, about one inch long, and purple or with whitish wings and keel. Sand pea grows on the coastal strand from extreme northern California to Alaska and also near the Great Lakes. Flowering is from May to July. Two introduced annual species with only two leaflets per stem are Tangier pea *(L. tingitanus),* which is not at all hairy, and the cultivated sweet pea *(L. odoratus),* which is hairy. Both the Tangier pea and the sweet pea escape along the coast.

The crowberry family (Empetraceae) is small and has low, heathlike, evergreen shrubs with slender, freely branched stems and rigid, narrow leaves. **BLACK CROWBERRY *(Em-***

Sand pea

petrum nigrum) has prostrate or spreading stems to about one foot long. The leaves are one-fourth inch long and thickened. The minute, purplish flowers usually have three sepals and petals and are solitary in the axils. The black or red berry has several nutlets. Blooming in spring, black crowberry occurs in dense beds in rocky places on sea bluffs from extreme northern California to Alaska.

In the sumac or cashew family (Anacardiaceae), a woody group often with poisonous or acrid sap, we find poison-oak *(Toxicodendron diversilobum)*, cashews *(Anacardium* spp.), mangoes *(Mangifera* spp.), and pistachios *(Pistacia* spp.). **LEMONADEBERRY *(Rhus integrifolia)*** is a rounded, or, near the sea, wind-pruned shrub three to nine feet tall, with reddish stout twigs and flat leathery entire or toothed leaves one to two inches long. The flowers are in dense clusters, more or less pinkish or rose, and quite small. They produce flattened, sticky, acidic fruits almost half-an-inch in diameter. This plant is common on sea bluffs and in coastal canyons from Santa Barbara County, California, southward.

Lemonadeberry

In the mallow family (Malvaceae), the stamens usually form a more or less complete tube around the several styles. **CHECKER MALLOW (Sidalcea malviflora)** is a perennial that has widely spreading rootstocks and stems one-half to two feet tall. The basal leaves are mostly entire, and those along the stem are deeply lobed. The rose or pink flowers are borne along an elongate stem and are one to two inches across. It occurs in grassy, often damp places from the coast inland and from southern Oregon to the Mexican border. Flowers appear in spring and summer.

Checker mallow

Malva rosa, or island mallow

Another member of the mallow family is **MALVA ROSA,** or **ISLAND MALLOW,** *(Lavatera assurgentiflora),* apparently originally native on the islands off the California coast, planted on the mainland, and now abundantly escaped, at least in the south. It is a shrub or small tree, three to 12 feet high, and has large leaves two to six inches wide. The petals are one to two inches long and rose with darker veins. The illustration shows the stamens of several lengths forming a tube around the central styles. Flowers are from March to November.

CALIFORNIA LOOSESTRIFE *(Lythrum californicum)* is of the loosestrife family (Lythraceae), to which also belongs crepemyrtle *(Lagerstroemia indica),* a tree commonly grown in warm regions. Erect and somewhat woody at the base, California loosestrife grows to over four feet high and has pale green, narrow leaves up to about an inch long. The flowers

California loosestrife

have purple petals one-fourth inch long. It is found in moist places near the coast and away from it, from northern California southward, and bears flowers from April to October. More common is the smaller, nonwoody hyssop-leaved loosestrife (*L. hyssopifolium*), which has shorter, rose to pink petals and is found in similar habitats throughout California.

The frankenia family (Frankeniaceae) is small and represented on the West Coast by two species; one of which is **ALKALI-HEATH** *(Frankenia salina)*. It is bushy, somewhat woody at the base, and up to one foot high, and the lower leaves are united in pairs by a membranous base. The scattered flowers are small and pinkish, and the seedpod is linear. Alkali-heath is found in salt marshes and on moist beaches from Marin and Solano Counties, California, to Baja California. It flowers from June to October.

California-fuchsia, or zauschneria

In the evening-primrose family (Onagraceae) are several characteristic western plants, among them **CALIFORNIA-FUCHSIA,** or **ZAUSCHNERIA,** *(Epilobium canum)*. A grayish perennial that is somewhat woody at the base and has narrow, clustered leaves and brilliant tubular scarlet flowers about an inch long, it commands attention in late summer and early fall. The ovary is inferior, that is, below the flower, and there are four two-lobed petals and four sepals. This plant grows in rocky places near the coast and inland throughout much of California and Oregon.

The clarkias are a conspicuous western group in the evening-primrose family. Once divided into two genera, *Clarkia* and *Godetia,* they are now all included in *Clarkia.* They are annual plants that are usually branched, often exfoliating on the lower stems, and have a large, narrow, inferior ovary (attached below the sepals and petals) and, in maritime species, usually four entire petals. A clarkia found near the coast is **FAREWELL-TO-SPRING** *(Clarkia amoena)*. Erect or sprawling, it is one to three feet tall, and its leaves can be over two inches long. The petals are an inch or more long, fan shaped to obovate, pink or laven-

Farewell-to-spring

der to white above, often pink or lavender at the base, and mostly penciled or blotched in the center with bright red. This clarkia is found on bluffs and slopes near the sea along the California coast north of San Francisco Bay. In similar places from Marin and Alameda Counties southward to Santa Clara and Santa Cruz Counties, California, grows ruby chalice clarkia *(C. rubicunda)*, a similar plant with lavender to somewhat pinkish petals and usually a bright red base but no central blotch. Both species bloom from May to July or August.

DAVY'S GODETIA *(Clarkia davyi)* is prostrate or nearly so and has simple or branched stems and rather crowded leaves less than an inch long. The flower is about half-an-inch long or less and has no spots or blotches. It is found in mostly sandy places along the coast from Del Norte to San Mateo Counties, California, and flowers in June and July.

Another member of the *Clarkia* genus is **ELEGANT CLARKIA** *(Clarkia unguiculata)*. It is erect and one to three feet high and has leaves to over two inches long. The petals, narrowed below into a slender claw, vary in color from lavender pink to salmon, purplish, or dark red purple. The species is common on dry, often shaded slopes but occurs at the coast even on back beaches. It ranges the entire length of California and blooms in May and June. It was introduced early to Europe

Davy's godetia

Elegant clarkia

where double and many other horticultural forms were developed and now appear in gardens.

RED RIBBONS (*Clarkia concinna*) grows in openings along the coast from the San Francisco Bay Area northward to Humboldt County. Like other *Clarkia* species, it is an annual, but it flowers from May to July long after most other native annuals have gone to seed. The petals are deeply divided into three lobes, and the bright, deep pink flowers are a welcome sight during a season when so much of the coastal landscape has turned brown in the summer drought. Red ribbons reaches two feet in height and favors road banks and other slightly disturbed areas.

Red ribbons

SALT MARSH–FLEABANE (Pluchea odorata), of the sunflower family (Asteraceae), is a glandular-hairy herb one to three or more feet tall and has toothed leaves two to four inches long. The flowers are in large terminal clusters made up of small heads of purple tubular disk flowers. This plant is found in marshy places along the coast and away from it, from San Francisco Bay southward, and also on the Atlantic Coast. It blooms from July to November.

The thistles are also in the sunflower family and have numerous, minute, elongated disk flowers that are compacted into a head surrounded by an involucre. A well-distributed coastal species is **INDIAN THISTLE (Cirsium brevistylum).** Ranging from British

Indian thistle

Milk thistle

Columbia to southern California, it is a short-lived perennial three to five feet high, more or less crisp arachnoid (cobwebby), and leafy to the top. The leaves of this species are more or less loosely woolly beneath and up to six inches long. The flowers are dull purple red.

MILK THISTLE *(Silybum marianum)* was introduced from Europe years ago and has become a common pasture weed, but it has established itself also on back beaches and dunes. It is erect, branched, and three to five feet tall and bears large, leathery, wavy or crisped leaves mottled with white blotches. The elongated purple disk flowers are in hemispherical heads with several spine-tipped leaflike bracts below, as in a true thistle. Flowers bloom from May to July and produce numerous seeds that are carried through the air by the downy tuft on each seed.

Among the most attractive members of the lily family (Liliaceae) are the mariposa lilies, or star-tulips, (*Calochortus* spp.), which have a flower plan of three instead of the four and five structures of broad-leaved plants.

Large-flowered star-tulip

LARGE-FLOWERED STAR-TULIP *(Calochortus uniflorus)* has an underground bulb, few linear leaves, and one to five lilac flowers. The petals are to about an inch long and often have a purple spot above the basal gland. This plant grows in low, wet, often alkaline places from Monterey County, California, to southwestern Oregon and blooms from April to June.

The flowers of **DWARF BRODIAEA *(Brodiaea terrestris* subsp. *terrestris),*** a member of the lily family, are often borne on the

Dwarf brodiaea

ground and individually may appear to emerge directly from the ground. Their bright blue color makes them easy to find in sandy, open areas along the coast from San Luis Obispo County, California, to Coos County, Oregon. By the time the flowers are produced, the leaves may have long since withered. The species survives the dry season by an underground bulb.

Douglas' iris

The most common iris near the coast is **DOUGLAS' IRIS *(Iris douglasiana)***. A member of the iris family (Iridaceae), this species forms heavy, sturdy clumps with leaves to three-fourths of an inch wide and flower stalks to over two feet tall. Flower color varies from pale cream to lavender or deep red purple. The range is from Santa Barbara County, California, to Oregon, and the flowering season is from March to May.

BLUE-EYED-GRASS *(Sisyrinchium bellum)* is not a grass but a member of the iris family. Plants are erect and up to two feet tall, and when they are not in flower, the leaves are very grasslike in appearance. Blue-eyed-grass is widespread all along the Pacific Coast, and, although individual flowers are short lived, plants flower over a long period of time and form colorful patches in coastal grasslands. The species is able to tolerate a certain amount of disturbance, and because plants are not eaten by cattle, light grazing may favor their spread. A handsome dwarf race of blue-eyed-grass from coastal California is available in the nursery trade and makes an attractive addition to rock gardens.

Blue-eyed-grass

Larkspurs (*Delphinium* spp.) are in the buttercup family (Ranunculaceae). **ANDERSON'S LARKSPUR *(Delphinium andersonii)*** has broad net-veined leaves and five-parted flowers. It has a small, shallow, underground cluster of tubers, stems one to two feet high, and soft, spreading white hairs. The leaves are usually one to two inches wide, and the flowers are few, with deep blue sepals half-an-inch long and smaller, paler petals. This larkspur is found in open places on bluffs above the ocean from Mendocino County, California, to British Columbia and bears flowers from March to May.

Anderson's larkspur

The Pacific states' lupines (*Lupinus* spp.), in the pea family (Fabaceae), are a diverse group with perhaps 100 species represented: some annual, some matted, some perennial, and some shrubby. Along the immediate coast, an interesting species is **SEASHORE LUPINE (*Lupinus littoralis*),** a slender-stemmed perennial that has spreading hairs and is mostly prostrate or nearly so. The pea-shaped flowers are about half-an-inch long and blue or lilac, the roots are yellow, and the leaf stems are one to two inches long. It is found along the immediate coast from northern California to British Columbia.

Closely related to seashore lupine *(Lupinus littoralis)* is **LINDLEY VARIED LUPINE *(L. variicolor)*,** which is also coastal but

its roots are not yellow and its leaf stems are two to four inches long. It has slender, more or less prostrate stems; somewhat hairy, green foliage; and seven to nine leaflets more or less silky beneath. The flowers are about half-an-inch long and are yellow, whitish, pink, purple, or blue. It ranges from Humboldt County, California, southward to San Luis Obispo County.

Lindley varied lupine

Sky lupine

SKY LUPINE *(Lupinus nanus)* is an erect annual with rich blue petals except for the white spot on the banner (upper petal). It is mostly a plant of the interior, covering great areas of grassy fields with its lovely blue, but it comes down to the coast on bluffs and beaches, where it can be found from Mendocino to Los Angeles Counties, California, particularly in April and May. It attains a height of one-half to one-and-a-half feet, and the whorled flowers are about half-an-inch long.

Still another lupine is **RIVER-BANK LUPINE** *(Lupinus rivularis)* of wet or sandy places along the immediate coast from Mendocino County, California, to British Columbia. It is a more or less hairy perennial herb and one to three or more feet tall and has petioles one to two inches long and leaflets one to one-and-a-half inches long. The numerous, somewhat whorled flowers are blue, purplish, or almost reddish, and half-an-inch long.

Riverbank lupine

Bush lupine

Along the coast, the most abundant shrubby lupine is **BUSH LUPINE** *(Lupinus arboreus),* a silky bush two to six feet high, with short leafy branches and six to nine broad leaflets. The inflorescence of whorled flowers is two to six inches long. Remarkable in having at least two color forms, one yellow, the other bluish or sometimes whitish or lilac, it is found from Ventura County, California, to Lane County, Oregon. Another shrubby lupine is silver lupine *(L. albifrons).* Similar to the bush lupine, its flowers are violet to lavender, never yellow, and it occurs inland as well as near the coast. In silver lupine, the upper banner petals have hairs on the backs, and the keels have fine hairs along the upper edges from the middle to the tip, whereas bush lupine has glabrous banner petals and fine hairs along the whole length of the upper edges of the keels.

GIANT VETCH *(Vicia gigantea),* a vinelike herb, is a stout-stemmed perennial, somewhat hairy, and two to three feet high and has tendril-bearing leaves made up of many leaflets. The numerous flowers are reddish purple, half-an-inch long, and occur on only one side of the stem. It is an inhabitant of moist places near the coast but is rarely out on the open beach. It ranges from San Luis Obispo County, California, northward to Alaska.

Giant vetch

In the buckthorn family (Rhamnaceae), the most conspicuous western genus is *Ceanothus,* which falls into two main groups: California-lilacs and buckthorns. California-lilacs have thin, deciduous stipules, alternate leaves, and flowers in terminal clusters; buckthorns have stipules with thick, corky persistent bases, leaves often opposite each other, and flowers mostly in lateral clusters. A coastal example of California-lilac is **CARMEL CEANOTHUS *(Ceanothus griseus),*** which has green, angled branchlets; broad, dark green shining leaves; and violet blue flowers in dense clusters one to two inches long. This plant occurs in central and northern California. Other similar blue-flowered or white-flowered species are found along the California coast and as far north as Coos and Curry Counties, Oregon.

Carmel ceanothus

DWARF CEANOTHUS *(Ceanothus dentatus)* is another California-lilac. A densely branched evergreen shrub that attains a height of two to five feet, its branchlets are hairy, and the small, toothed leaves are crowded, clustered, and rather narrow. The small, deep blue flowers are arranged in clusters up to

Dwarf ceanothus

two inches long. It occurs in sandy and gravelly places near the coast from Santa Cruz to San Luis Obispo Counties, California. Other species extend much farther north.

In the violet family (Violaceae), our western violets (*Viola* spp.) come in several colors, and a bluish one of scrubby bluffs and banks along the coast is **MARSH VIOLET** *(Viola palustris)*.

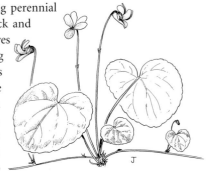

It is a smooth, creeping perennial with a slender rootstock and somewhat roundish leaves two to four inches long and one to two inches wide. The flowers are lilac to almost white with some darker veining, and the lateral petals are somewhat hairy. It ranges from Mendocino County, California, to Alaska and also occurs on the Atlantic Coast and in Europe. It blooms largely from May to July.

The carrot family (Apiaceae) is a large family usually characterized by its aromatic nature due to its contained oils. Some members do not have the well-developed umbels usually associated with this family, such as **PRICKLY BUTTON-CELERY,** also known as **PRICKLY ERYNGIUM,** or **COYOTE THISTLE,** *(Eryngium armatum)*. It is a spiny plant with stems up to a foot or more long. The minute flowers can be creamy, purple, or white and are arranged in tight heads surrounded by bluish or yellowish leaflike bracts. It is found in moist places that dry with the advancing season and occurs near the coast from Santa Barbara to Humboldt Counties, California.

WESTERN MARSH-ROSEMARY, or **WESTERN SEA-LAVENDER,** *(Limonium californicum)* is in the leadwort family (Plumbaginaceae). It has a stout stem and is somewhat woody at the base.

Prickly button-celery, prickly eryngium, or coyote thistle

Western marsh-rosemary, or western sea-lavender

The leaf blades can be up to eight inches long, and the flowering stems are up to almost two feet high. The flowers are small and have whitish sepals and pale violet petals. Several European relatives of the plant are cultivated for their flower clusters, which become papery with age and can be used for dry bouquets. They are cultivated along the coast and sometimes escape. The native species is found on beaches and along the upper margins of salt marshes along most of the California coastline.

The phlox family (Polemoniaceae) is for the most part not maritime, but on the coastal strand of central California are a number of subspecies of *Gilia capitata* such as **DUNE GILIA** *(G. capitata* subsp. *chamissonis)*. This gilia is a tufted annual, glabrous to glandular or woolly, one to three feet tall, and branched at the base or above and has dissected leaves. The flowers are in close clusters or heads, with a calyx of five narrow sepals grown together and a funnel-shaped corolla of five bluish, narrow to oval petals. Flowers appear from spring to early summer.

Dune gilia

Eriastrums (*Eriastrum* spp.), close relatives of the genus *Gilia*, are other members of the phlox family and are distinguished by unequal calyx lobes and flower heads underlain by leaflike bracts. One of the coastal species is **PERENNIAL ERIASTRUM**, or **MANY-LEAVED ERIASTRUM**, *(Eriastrum densifolium)*, an erect, much-branched perennial to about one foot high. The leaves are linear and entire or divided into linear lobes. The trumpet-shaped blue corollas are about an inch long. The typical form grows in sandy places along the coast from Monterey County, California, to Santa Barbara County, and a closely related subspecies occurs in Orange County.

The lennoa family (Lennoaceae) is a small family of fleshy herbs that are parasitic on roots of other plants. Its members

Pholisma

lack chlorophyll but turn more or less brown when dry. A coastal representative is **PHOLISMA *(Pholisma arenarium)*.** The part above ground is four to eight inches tall and clumped and has a whitish stem that ages brown. Pholisma has bractlike leaves and numerous purplish flowers with white borders. It is occasional in sandy places from San Luis Obispo to San Diego Counties, California. The flowers appear for the most part from April to July.

COMFREY *(Symphytum asperum)* is a member of the borage family (Boraginaceae). The flowers are usually arranged in coiled branches, and the ovary produces four one-seeded nutlets. Comfrey is a native of Asia and is naturalized near the coast in parts of northern California. It has a deep root, stems with recurved, hooklike hairs, and bluish flowers about half-an-inch long.

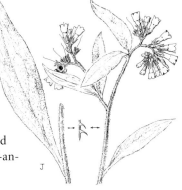

Another family with flowers in coiled branches but with undivided ovaries so that the seeds are borne in small pods is the waterleaf family (Hydrophyllaceae), of which *Phacelia* is a large genus. **BOLANDER'S PHACELIA *(Phacelia bolanderi)*** is a perennial from a root crown, and the few stems are up to two or even three feet tall and mostly hairy and have broad leaves two to four inches long. The open, spreading corolla is lilac to pale blue. This species occurs along the immediate coast from Sonoma County, California, to Oregon but is sometimes found inland as well. Flowers appear from May to July.

BABY BLUE-EYES *(Nemophila menziesii)* is one of the earliest wildflowers to bloom in spring along the California coast and

Baby blue-eyes

in interior areas. The weak stems of these annual plants scramble among other plants, going unseen until their delicate blue flowers open to signal that spring is on its way. This species, another member of the waterleaf family, is variable throughout its range. One coastal variety *(N. menziesii* var. *atomaria)* has white flowers speckled with black dots, another *(N. menziesii* var. *integrifolia)* has blue flowers that are black dotted in the center, and still another *(N. menziesii* var. *menziesii)* has blue flowers with white centers. This last variety has the largest flowers of all, with petals nearly an inch long. Baby blue-eyes is also a long-time favorite bedding plant in the United Kingdom.

Members of the mint family (Lamiaceae) have united petals that form an upper lip and a lower lip. The stems are often square, the leaves are opposite, and the fruit consists of four one-seeded nutlets. Usually the plants are highly aromatic, as in spearmint *(Mentha spicata)*, bergamot *(M. citrata)*, and pennyroyal *(M. pulegium)*. **BLACK SAGE *(Salvia mellifera)*** is a many-stemmed shrub three to six feet high and has green

leaves with impressed veins and heads of pale blue to lavender or whitish flowers about half-an-inch long. It may grow on bluffs overlooking the ocean from Contra Costa County, California, southward.

Another family with two-lipped corollas is the figwort family (Scrophulariaceae). A genus of annuals sometimes called Chinese houses (*Collinsia* spp.) has the middle lobe of the lower lip folded into a small boatlike structure, or keel, that contains the stamens and pistil. 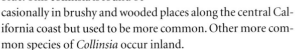 **SAN FRANCISCO COLLINSIA** *(Collinsia multicolor)* is one to two feet tall and often somewhat sticky above and has paired leaves and flowers three-fourths of an inch long. The upper lip is whitish, and purple spotted near the base, whereas the lower lip is violet blue. This collinsia is found occasionally in brushy and wooded places along the central California coast but used to be more common. Other more common species of *Collinsia* occur inland.

In the same family is **SNOW QUEEN** *(Synthyris reniformis)*, found in rich, coniferous forests from Marin County, California, to Washington. A rather hairy perennial, its leaves are at the base of the plant, its stems are to about six inches high, and its blue flowers are about one-third inch long. The capsule is two lobed.

Nightshades (*Solanum* spp.), in the nightshade family (Solanaceae), have, in general, wheel-shaped flowers and strong-smelling herbage when crushed. **BLUE WITCH** *(Solanum umbelliferum)* is not primarily coastal but grows on the

Snow queen

Blue witch

California-aster

dunes of Morro Bay in San Luis Obispo County, California, and at other points along the California coast. The bluish purple flowers are half-an-inch or more across and are borne in small, open clusters. The fruit is a berry that resembles a small, green tomato. Some flowers appear during most of the year.

CALIFORNIA-ASTER (*Lessingia filaginifolia* var. *californica*) is a member of the sunflower family (Asteraceae) and has many minute flowers in heads with layers of leaflike bracts below the flower heads. This species is a close relative of the true aster but is distinguished by having a brushlike microscopic appendage on the style of the pistil. It is perennial, more or less permanently white woolly, and has violet purple to lilac pink petal-like ray flowers and yellow disk flowers. Along the coast it has two forms: one with narrow oblanceolate leaves occurring from the Golden Gate to Monterey, the other with spatulate to obovate (egg shaped with the broad end up) leaves ranging from Marin County, California, to Coos County, Oregon.

Seaside daisy

SEASIDE DAISY *(Erigeron glaucus)* is a low perennial with stems from four to 16 inches high. The leaves are entire or somewhat toothed, mostly basal, and three to six inches long. The heads terminate long branches and have very numerous (perhaps up to 100) pale violet to lavender petal-like ray flowers about half-an-inch long and many yellow disk flowers. Common on coastal bluffs and beaches from Clatsop County, Oregon, to central California, this plant blooms from April to August.

SURF-GRASS (Phyllospadix torreyi) and **EEL-GRASS (Zostera marina)** belong to the eel-grass family (Zosteraceae), a group of seed-bearing plants that look like seaweeds but are not.

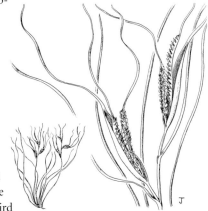

Both species grow submerged in shallow water in bays near the shore and are tossed up on the sand in times of storm. Both have two rows of leaves and minute, greenish petal-less flowers arranged on one side of a flattened stem. In eel-grass, the leaves are one-twelfth to one-third inch wide, and in surf-grass, they are less than one-twelfth inch wide. The former ranges from San Diego to Alaska and in Eurasia, and the latter ranges from Baja California northward.

Another plant that is scarcely a wildflower but is a flowering plant and attracts attention by its odd appearance is **SEASIDE ARROW-GRASS (Triglochin maritima)**. A member of the arrow-grass family (Juncaginaceae), it is a marsh herb and densely tufted. It has stiff, narrow leaves and terminal flowering spikes one to two feet high. The minute, greenish flowers have six perianth segments and produce clusters of three or six one-seeded fruits. This species of coastal salt marshes ranges from San Francisco Bay northward, but other related species are found as far southward as Baja California.

CALIFORNIA CORD GRASS (Spartina foliosa) is a true grass, having hollow stems and swollen nodes. In the grass family (Poaceae), this species extends intermittently along the sandy and marshy shore from Del Norte County, Cal-

Seaside arrow-grass

California cord grass

ifornia, to Baja California, but related species go far to the north. This plant is a coarse perennial up to several feet tall. It has strong, creeping rootstocks, which makes it important in expanding the margins of salt marshes. The inflorescence is up to a foot long and is somewhat cylindrical and made up of numerous spikelets, each with two leaflike basal bracts, or glumes, and a single, petal-less floret with stamens and a pistil.

EUROPEAN BEACHGRASS (*Ammophila arenaria*) is a common grass that was imported from Europe to stabilize shifting sand dunes along the Pacific Coast. It has served this purpose very well because it forms a dense, large turf. But native plant species cannot compete with this grass, so where you encounter it, you will see few natives. In addition, European beachgrass burns very easily and regenerates vegetatively after fires, whereas the few native plants that may survive in its company do not reappear after fires.

European beachgrass

Among other grasses of coastal salt marshes with striking appearances is **SHOREGRASS (*Monanthochloe littoralis*),** a spreading, wiry-stemmed perennial with clusters of short, awl-shaped leaves and short, erect branches. The flower-producing spikelets are scarcely evident, and the staminate and pistillate flowers are on separate plants. This grass occurs from Santa Barbara County, California, to Baja California, and also in Texas, Florida, Cuba, and Mexico.

Associated with shoregrass (*Monanthochloe littoralis*) in salt marshes and often forming large patches is **SALTGRASS *(Distichlis spicata),*** which ranges from Oregon to southern California. It grows from strong, creeping or deeply running rootstocks and has two-ranked leaves four to eight inches long. The flowering spikelets are evident in dense clusters and are more or less green or sometimes purplish. Some forms of this grass are found in salty places inland, even in the desert.

Saltgrass

The bluegrasses (*Poa* spp.) have two to eight florets in a single spikelet and a basal pair of glumes. **SAND DUNE BLUEGRASS *(Poa douglasii),*** the common *Poa* of the beach, is a low, tufted grass spreading by deep-seated rhizomes, with aerial runners up to two or three feet long. The rigid stems are six to 16 inches long, and the leaf blades are stiff and inrolled (see the illustration). The three- to nine-flowered, pale tawny spikelets are in dense clusters one to two inches long. Sand dune bluegrass ranges from California's Channel Islands to Puget Sound.

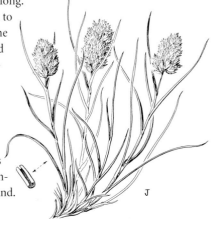

One more unusual and interesting grass of salt marshes and coastal strand is **SICKLE GRASS (Parapholis incurva)**. In contrast to the grasses mentioned above, this grass is an annual with slender, cylindrical, curved spikes in which the inconspicuous one- or two-flowered spikelets are embedded. It occurs from Oregon to southern California and on the Atlantic Coast but is native to Europe.

Sickle grass

The sedge family (Cyperaceae) is close to the grasses (Poaceae), but sedges have three-sided stems that are usually not hollow, and they do not have swollen, hard nodes. The florets are arranged in spikelets, as in the grasses, but lack the two basal glumes. Cotton-grass (*Eriophorum* spp.), tule and bulrush (*Scirpus* spp.), papyrus (*Cyperus papyrus*), and sedge (*Carex* spp.) are in this family. Along our western shore are several members of this family, and two examples are presented here. One is **BULRUSH (Scirpus maritimus),** a perennial with horizontal tuber-forming rhizomes and erect, sharply triangular stems one-and-a-half to four-and-a-half feet tall. At the summit is a tuft of unequal leaves below a cluster of ovoid spikelets that are up to an inch long and have pale gray to brownish scales. Each floret has several bristles that represent the modified perianth. Bulrush grows in salt marshes of California and Oregon, as well as on the Atlantic Coast and to South America.

Bulrush

A second sedge is **SAND DUNE SEDGE (Carex pansa),** a tufted plant growing from long, creeping rootstocks. It attains a height of about one foot and has few to several crowded spikelets with the stamens at the summit. It is found on beaches and coastal dunes from northern California to Washington.

Another family that resembles grasses and sedges in its small flowers is the rush family (Juncaceae), but each flower has six small, modified, greenish perianth parts, as well as stamens and pistils, and looks like a small lily in structure. **SPINY RUSH (Juncus acutus subsp. leopoldii)** is a perennial that forms large tufts two to four feet high

Fairy bells

and has stout, stiff stems and small clusters of two to four flowers up to one-sixth inch long. This plant inhabits coastal salt marshes from San Luis Obispo County, California, to Baja California.

A plant with larger and more typical-looking flowers is **FAIRY BELLS *(Disporum smithii),*** of the lily family (Liliaceae). A perennial herb from slender rootstocks, it sends up branched stems to over two feet long that have several broad leaves and one to five mostly whitish flowers in a cluster. The perianth segments are up to an inch long. The fruit is a light orange to red berry. The species occurs in moist, shaded woods and reaches the shore at intervals from Santa Cruz County, California, to British Columbia. It flowers from March to May.

SLIM SOLOMON'S SEAL *(Smilacina stellata)* is at home in coastal forests, as well as on grassy slopes along the coast from California to British Columbia. The species name *stellata* refers to the starlike appearance of the three to nine small white flowers. These are followed by dark red or blackish fleshy fruits. The plant has slender, creeping underground

Slim solomon's seal

stems and often forms large colonies of its graceful, arched aerial stems that hide the ground. Plants vary in size, and stems can reach two feet in length.

Also in the lily family, with its six stamens, superior ovary, and three inner and three outer, usually petal-like perianth segments, is **FRAGRANT FRITILLARY (Fritillaria liliacea)**. A bulbous plant, with a stem four to 14 inches high and leaves just above ground level, it has one to five bell-shaped whitish flowers half-an-inch or more long with green striations. This increasingly rare plant is found in heavy soil near the coast and ranges from Sonoma to Monterey Counties, California. The flowers appear from February to April.

Fragrant fritillary

FALSE LILY-OF-THE-VALLEY (Maianthemum dilatatum) is a low, perennial herb with a creeping rootstock. It forms large patches in moist shaded places and grows down on to the actual beach in such spots. The stems are six to

False lily-of-the-valley

Star-lily

15 inches high, and the small white flowers are followed by red berries. The range is from Marin County, California, to Alaska and Idaho. Flowering is from May to June.

Death-camases (*Zigadenus* spp.) are so named because the leaves and bulbs of these members of the lily family contain alkaloids that are toxic to humans and livestock if they are eaten. **STAR-LILY** *(Zigadenus fremontii)* is two to three feet tall and the most beautiful and attractive of the star-lilies, or death-camases. It grows from near the Mexican border northward to southwestern Oregon. It often grows with the blue-flowered true camas *(Camassia quamash)*, whose bulbs are highly prized as food by various Native American groups in western North America. In flower, the star-lily and camas are easily distinguished, but when in leaf only, they are hard to tell apart. Star-lily is one of the earliest coastal wildflowers to bloom in spring and is especially vigorous when growing on a recently burned site.

FRINGED CORN-LILY, or **FRINGED FALSE-HELLEBORE,** *(Veratrum fimbriatum)* is a bold, tall lily. It grows to over six feet tall and can be found in wet meadows of coastal Sonoma and Mendocino Counties, California. Other species of corn-lily occur eastward to the Rocky Mountains and northward to

Fringed corn-lily, or fringed false-hellebore

Alaska. The broad leaves clasp the stem and are traversed by conspicuous veins and longitudinal pleats. The attractive white flowers are borne in large masses and are individually characterized by their fringed petals. All corn-lilies contain alkaloids that are toxic to humans and livestock if the plant is ingested.

Also in the lily family is **GIANT TRILLIUM,** or **GIANT WAKE-ROBIN,** *(Trillium chloropetalum)*. With stout stems a foot or more in height, it bears three large, often mottled leaves and a single sessile flower with three petals from one-and-a-half to almost four inches long. The petals range from white to pink to a deep burgundy. There are several forms of this species that may appear near the shore from Santa Barbara County, California, northward to Washington and that bloom between February and May.

The elegant **WHITE TRILLIUM,** or **WESTERN TRILLIUM,** *(Trillium ovatum)* is a member of the lily family that can be found flowering as early as March. It may reach a foot in height and typically grows in moist, shady wooded areas and ranges along the coast from central California into British Columbia. When shading becomes too great, plants may cease flowering for decades, producing only leaves and no flowers. If fires or fallen trees produce light gaps, the plants begin to flower once again. The seeds of trilliums are dispersed by ants, who feed on oily structures attached to the seed coat. The pure, white flowers are long lived and in many plants turn deep pink with age. Resist the urge to pick trilliums because

White trillium, or western trillium

you will also remove the leaves and set the plant back. This species is practically impossible to grow in garden conditions, so enjoy it in the wild.

In the orchid family (Orchidaceae) is the summer-flowering **LADIES' TRESSES** *(Spiranthes romanzoffiana)*. It is frequent in wet meadows from California to southern Alaska, and the

Ladies' tresses

small, greenish white or yellowish flowers pale in comparison with gaudy greenhouse orchids from tropical regions. The name *Spiranthes* means "spiral flowers," and if you examine the flowering stalk you will see that the rows of flowers are arranged in a long spiral along the stalk. The common name probably originated at a time when the word "tress" referred to braided hair rather than to a lock of hair, and in this plant it refers to the braidlike arrangement of flowers. This orchid ranges from four to 18 inches tall.

Another member of the orchid family that is found on the immediate coast is **ELEGANT PIPERIA,** or **ELEGANT REIN ORCHID, (*Piperia elegans*).** It has fleshy leaves and a dense spike of greenish white flowers, each with a spur half-an-inch or longer. It can be expected on sea bluffs and in similar places from Monterey County, California, to Oregon and blooms from July to September.

Elegant piperia, or
elegant rein orchid

Yerba mansa, or lizard's tail

Members of the lily family (Liliaceae) and orchid family (Orchidaceae) are monocotyledons, one of the two main divisions of flowering plants. Monocotyledons have flowers built on a plan of three and parallel-veined leaves. Dicotyledons have broad, net-veined leaves, and the flowers are mostly on a plan of four or five. One dicotyledon, in the lizard's tail family (Saururaceae), is **YERBA MANSA**, or **LIZARD'S TAIL** *(Anemopsis californica),* Highly specialized, it has a cluster of many flowers that resembles a single one because the white petal-like parts around the base are actually modified bracts and the individual flowers are minute without sepals or petals but with six or eight stamens and a pistil. Yerba mansa grows in wet places from Oregon southward through California and then eastward to Texas.

WAX-MYRTLE *(Myrica californica)* is a tall shrub of the wax-myrtle family (Myricaceae). Another member of this genus in the northeastern states, bayberry *(M. pensylvanica)* produces the bayberry wax so famous for use in candles, and, as a matter of fact, our western species has a whitish wax that covers its small fruits. It is a large shrub up to 12 feet tall or more and

Wax-myrtle

evergreen and has shining leaves to almost four inches long and minute flowers in catkinlike clusters. This plant inhabits mostly canyons and moist slopes and grows at low elevations along the coast from the Santa Monica Mountains near Los Angeles northward to Washington.

Quite a contrast to wax-myrtle *(Myrica californica)*, but also with small, greenish flowers in catkins, is **HOARY NETTLE**, or **STINGING NETTLE**, *(Urtica dioica* **subsp.** *holosericea)*, a perennial herb in the nettle family (Urticaceae) that grows from underground rootstocks and has stems three to seven feet high. It is covered with bristly

Meadowfoam

hairs that are like small, glass bottles that break in the human skin and inject a small quantity of a stinging fluid. The small flowers are green and have a deeply parted calyx but no petals. It occurs widely in low, damp places and can be found along the shore edge as far north as Washington.

MEADOWFOAM *(Limnanthes douglasii)* is in the meadowfoam family (Limnanthaceae). A low, annual herb with alternate, pinnately divided leaves and solitary, three- to six-parted flowers, meadowfoam can form great masses in low places that are moist in spring. It is not primarily a coastal plant but occurs there occasionally, especially in the yellow-flowered form *(L. douglasii* subsp. *sulphurea)* on Point Reyes Peninsula, Marin County, California.

The ocean beaches have a number of conspicuous plants in the fig-marigold family (Aizoaceae), such as **CRYSTALLINE ICEPLANT** *(Mesembryanthemum crystallinum)*. It is an annual and usually has broad, alternate leaves, the surfaces of which are covered with shining, colorless, conspicuous projections. It is very succulent, prostrate, much branched, and variously

Crystalline iceplant

hued. The flowers are somewhat less than an inch in diameter and white to reddish. Crystalline iceplant occurs in sandy or saline places from Monterey County to Baja California. Another annual species with leaves largely alternate, but semi-rounded instead of flat, is slender-leaved iceplant (*M. nodiflorum*).

NEW ZEALAND SPINACH *(Tetragonia tetragonioides)* is a different-looking member of the fig-marigold family. It has many spreading branches and triangular leaves one to two inches long. The flowers are solitary in the leaf axils and greenish and have short spreading sepals and no petals. The horned fruit is hard and one-third inch long. An introduced annual originally from Southeast Asia and Australia, it has become naturalized along our beaches and near salt marshes from Oregon southward.

The buckwheat family (Polygonaceae) is remarkable among the broad-leaved plants because its flowers are in a plan of three. One of the conspicuous western genera is *Chorizanthe*,

New Zealand spinach

which is represented by **MONTEREY SPINEFLOWER *(Chorizanthe pungens)*.** A more or less prostrate annual, it has basal leaves, grayish-hairy stems up to a foot long, and dense, head-like clusters of minute flowers with six-parted greenish perianths, each segment of which ends in a recurved spine. Found in sandy places along the coast from Monterey to San Francisco, spineflowers have become increasingly rare because of coastal development and invasive nonnative plants.

Monterey spineflower

Wild buckwheats (*Eriogonum* spp.) differ from *Chorizanthe* spp. because they do not have spine-tipped perianth segments. Among the wild buckwheats are many quite attractive species, even though their flowers are very small. **LONG-STEMMED BUCK-WHEAT (*Eriogonum elongatum*)** is a perennial herb, whitish woolly throughout, and leafy below and has long, leafless branches up to two or four feet tall. The small white or pinkish flowers are in short, cylindrical, cuplike structures called involucres. This wild buckwheat can be found in rocky places along the coast from Monterey County, California, to northern Baja California, as well as farther inland, and bloom from August to November.

COAST BUCKWHEAT (*Eriogonum latifolium*) is in one of the most diverse and widespread groups of wild buckwheat. This species is woody and densely leafy at the base, with white-woolly leaves at least on the undersurface. The stout, leafless flowering stems are usually hairy, at least near the top, and usually fork. They may reach a height of two feet, and at their summits appear the small white to rose flowers about one-eighth inch long with three outer and three inner petal-like perianth segments. Coast buckwheat is found in sandy places along the coast from San Luis Obispo County, California, to Oregon.

Dune buckwheat

Another coastal wild buck-wheat is **DUNE BUCKWHEAT** *(Eriogonum parvifolium)*. It is woody, has almost prostrate branches, and is thinly woolly, especially on the undersides of the leaves. Common on bluffs and dunes along the coast, it ranges from Monterey to San Diego Counties, California, and bears white flowers tinged with pink. Some flowers can be found through most months of the year.

A fourth coastal buckwheat is **GRAY COAST BUCK-WHEAT** *(Eriogonum cinereum)*. It is woody, freely branched, and up to three or four feet tall. It is some-what woolly and has larger leaves than dune buck-wheat *(E. parvifolium)*. The clusters of whitish to pinkish flowers appear between June and December, and the individual flowers are about one-eighth inch long. This species is found on beaches and bluffs along the coast and ranges from Santa Barbara to Los Angeles Counties, California.

BEACH KNOTWEED *(Polygonum paronychia)*, also of the buckwheat family, is rather a remarkable species and quite different from common knotweed *(P. arenastrum)*, the backyard weed introduced from Europe. Beach knotweed is a more or less prostrate native perennial from large, woody rootstocks and much branched and has papery sheaths at the nodes and inrolled leaves, as shown in illustration. The flowers are small and white to pink with green midveins. It inhabits the coastal strand from Monterey County, California, to British Columbia.

Beach knotweed

Another *Polygonum*, **HIMALAYAN KNOTWEED** *(Polygonum polystachyum),* is included here because, although not widespread, it is very conspicuous where it does occur, growing to a height of three or four feet. It has leaves four to eight inches long and large, terminal clusters of small, white flowers. It is found on vacant lots and in coastal marshes in the regions of Polk County, Oregon, and Fort Bragg and Eureka in northern California. This plant is an introduction from Asia and may well spread more widely in our cool, northern coastal region. The flowers appear from June to September.

WESTERN BISTORT *(Polygonum bistortoides)* is another member of the buckwheat family and is a perennial with a thick, horizontal rootstock and with several erect, slender, simple, smooth stems one to two feet high. Commonly thought of as an inhabitant of moist places in the high mountains, it grows also in coastal marshes from Marin County, California, to Alaska, and on the Atlantic Coast. In these northern cool climates, it sends up its compact spikes of small, white or pinkish flowers with six-parted perianths. It blooms from June to August.

Western bistort

Also in the buckwheat family is **WILLOW DOCK *(Rumex salicifolius* var. *crassus),*** which has trailing to ascending stems that are one to one-and-a-half feet long and leaves about three times as long as wide. It is perhaps the most common of several dock species that occur along the coast. The outer sepals of the whitish or greenish flowers in this species are only about one-twelfth inch long, and the three inner sepals, or perianth segments, form valves that cover the fruit and become one-quarter inch long, one of the sepals being almost covered by a large callosity, or seedlike growth. This plant occurs on coastal dunes and rocky, ocean bluffs in Los Angeles County and from Monterey County to Washington. It blooms from May to September.

Common in low places, both in the interior and near coastal marshes, is **MEXICAN-TEA *(Chenopodium ambrosioides)*** of the goosefoot family (Chenopodiaceae). Like most of that family it has small greenish flowers without petals. It is an ill-smelling plant, annual to perennial, with sprawling stems to a yard long and with toothed to lobed leaves one to four inches long. The terminal clusters of glandular flowers are quite conspicuous. It is widely spread on the Pacific Coast, naturalized from tropical America.

SPEARSCALE, or **FAT HEN,** *(Atriplex triangularis)* is another member of the goosefoot family. The *Atriplex* genus is commonly known as saltbush, and plants in this genus are usually covered with a scurfy coating of inflated, balloonlike hairs. They have separate staminate (male) and pistillate (female) flowers, the latter being below the former or on  separate plants and situated between two bracts (see the triangular drawing in the illustration). Spearscale, or fat hen, grows in salt marshes of the interior and the coast, where it can be found as far north as British Columbia.

Another saltbush is an erect shrub known as **BIG SALTBUSH** *(Atriplex lentiformis).* Also grayish and scurfy, it attains a height of three to eight feet and bears leaves one to two inches long. The pair of bracts below the pistillate flower is shown in the lower right-hand corner of the illustration, and it can be seen how the species in this group differ in their fruiting bracts. This plant is found near coastal salt marshes and on bluffs along the shore and inland, from San Francisco Bay to southern California.

Likewise inhabiting salt marshes and low alkaline places, and likewise in the goosefoot family, is a plant with jointed leafless stems, **PICKLEWEED,** or **SAMPHIRE,** *(Salicornia subterminalis)*. Ranging from the San Francisco Bay Area to Mexico, it is a perennial, but annual species often occur with it. The flowers are sunken in a cylindical fleshy stem, green in color, and an inch or more long. This group of plants occurs worldwide.

Pickleweed, or samphire

Yet another member of the goosefoot family is **WOOLLY SEA-BLITE** *(Suaeda taxifolia)*, also of coastal salt marshes and environs. Many plants of this family grow in saline places along the coast or inland, and, interestingly, many are found in similar habitats in the interior of Asia. This species is perennial, woody at the base, usually densely hairy, and up to almost five feet high. It is much branched and glaucous and has fleshy leaves up to an inch or more long. The small, greenish flowers have a five-parted calyx and no petals, as is characteristic of the family. Its distribution is from southern California to Baja California and in the Channel Islands.

A final member of the goosefoot family, and one which because of its inconspicuousness should perhaps not be included here, is **APHANISMA (Aphanisma blitoides)**. My feeling,

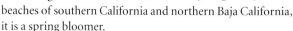

however, is that this book may help you use larger, more technical ones and that illustrations such as these may help place species that are difficult to handle by keys. This plant is a somewhat succulent annual and four to 20 inches tall and has leaves to an inch long and small, lens-shaped greenish fruits. Inhabiting coastal bluffs and beaches of southern California and northern Baja California, it is a spring bloomer.

Another inconspicuous but locally common maritime plant is **SALTWORT**, or **BEACHWORT**, **(Batis maritima)** of the saltwort family (Bataceae). Growing on coastal strand and in salt marshes of southern California, it occurs also on the Atlantic Coast, in the West Indies, and in South America. Prostrate or ascending, and woody at the base, the stems become a yard long and bear fleshy leaves half-an-inch

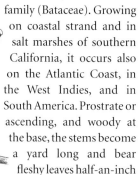

or longer. The flowers are crowded into catkinlike spikes with no calyx or corolla, the staminate flowers form four stamens, and the pistillate flowers form one ovary. The pistils coalesce to form a fleshy fruit.

Saltwort, or beachwort

We come now to the pink family (Caryophyllaceae), which has opposite leaves and often showy flowers. However, **CARDIO-NEMA,** or **SAND MAT,** *(Cardionema ramosissimum)* is far from conspicuous. It is a low, tufted, grayish perennial with short, branched, more or less woolly stems and papery scalelike stipules at the base of the leaves. In the illustration, the middle, upper drawing shows one of these stipules, which is two parted and has a stiff linear leaflike vein in the middle. The five-parted calyx has unequal sepals ending in short, pointed spines. The distribution is remarkable: sandy places along the coast from Washington to Baja California, Mexico, and Chile.

Beach starwort

Chickweeds, or starworts, (*Stellaria* spp.) are more representative of the pink family than is the genus *Cardionema*. A good example is **BEACH STARWORT (*Stellaria littoralis*)**, with its deeply two-cleft petals and three styles. A hairy perennial, it has forking stems up to 20 inches long, many oval leaves, and petals about one-fourth inch long. It grows on coastal strand and adjacent bluffs from San Francisco to extreme northern California and flowers from March to July.

Also of the pink family is **ARCTIC PEARLWORT (*Sagina procumbens*)**, a matted perennial. The prostrate delicate stems root at the nodes and are one to three inches long. The basal leaves are up to three-fourths of an inch long and bristle tipped. The flowers have mostly four sepals and petals, sometimes five, and are about one-twelfth inch long. The species is found on moist, shaded banks near the beach and on adjacent bluffs from Point Reyes, Marin County, California, northward to British Columbia and also on the Atlantic Coast. This plant is a native of Eurasia.

A last member of the pink family, with opposite, more or less fleshy leaves and papery, scalelike stipules, but with the white or pink petals not cleft as in beach star-wort *(Stellaria littoralis),* is **LARGE-FLOWERED SAND-SPURREY (*Spergularia macrotheca*).** The plant is perennial from a heavy-branched woody base and a fleshy root, and the stems tend to be prostrate and up to a foot long. This species of sand-spurrey inhabits sea bluffs and is found about salt marshes from British Columbia to Baja California. Somewhat different varieties grow in alkaline spots in interior valleys.

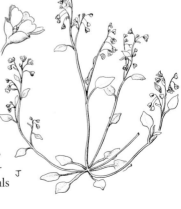

In the purslane family (Portulacaceae) is **DIFFUSE MONTIA (*Montia diffusa*).** A branched annual two to six inches high, its basal and stem leaves are alike and measure one to two inches long. The small, white flowers have the characteristic pair of fleshy sepals of the family and white or pinkish petals about one-sixth inch long. It comes down to the shore in wooded areas from Marin County, California, to Washington and is sometimes found almost to the edge of the sand. Flowers are from May to July.

Also in the purslane family is **MINER'S-LETTUCE (*Claytonia perfoliata*),** which is remarkable in its pair of united stem 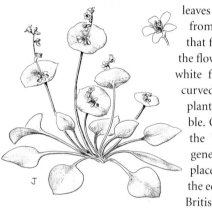 leaves that are so different from its basal leaves and that form a cup just below the flower cluster. The small, white flowers are often recurved in age. The whole plant is fleshy, and it is edible. Common in much of the West, the species is generally found in shaded places and may occur near the edge of the beach from British Columbia to Baja California. It flowers largely from February to May.

The mustard family (Brassicaceae), with its four-petaled flowers and two-chambered seedpod, usually has a biting or peppery sap. **SHARP-PODDED PEPPER-GRASS (*Lepidium oxycarpum*)** is a slender-stemmed little annual with leaves one to two inches long and small, white or green flowers with or without minute petals. The flat pod is one-eighth inch long and widens into two divergent lobes at the top. This plant is found in saline

Wild radish

flats and alkaline valley floors and occurs at the edges of coastal salt marshes of central California.

Another member of the mustard family is **WILD RADISH (Raphanus sativus)**. It bears rather showy, white or pink to purplish flowers with rose or purplish veins. A freely branched annual, it is erect and one to three feet tall and has prominently parted lower leaves. The fruits are characteristically narrowed between the seeds. It is a weed of vacant lots and fields, long ago naturalized from Europe, but it grows abundantly on back beaches and adjacent areas. It may cover large areas near the shore. Jointed charlock *(R. raphanistrum)*, with creamy yellow flowers with brown to reddish veins, generally grows with the wild radish. Technically, only the plants with pink to purplish flowers qualify as the true wild radish. The white-flowered plants are reported to be hybrids with a blocked color gene.

SEASIDE BITTER-CRESS *(Cardamine angulata)* is another member of the mustard family. It is a perennial growing from a slender, running rootstock, suberect, and one to two-and-a-half feet tall and has angularly lobed leaflets and white petals about half-an-inch in length. The spreading elongate seedpods are an inch or so long. It is a forest inhabitant from northern California to British Columbia and comes down to the shore in that cool northern area. Flowers appear largely in May and June.

Coast boykinia

The saxifrage family (Saxifragaceae) is represented by several plants, such as **COAST BOYKINIA** *(Boykinia occidentalis)*. A slender-stemmed perennial herb, it is erect, one to two feet high, with minute brown, gland-tipped hairs. The lower leaves are one to three inches wide, and the white flowers are one-eighth inch long. It is rather a dainty plant, usually of shaded springy places, and occasionally comes out to the shore. This plant is found from southern California to Washington.

Another member of the saxifrage family, and one that always intrigues me by its finely divided petals, is **COASTAL MITRE-WORT** *(Mitella ovalis)*. It is a low perennial to about one foot high and has recurved hairs and greenish petals. We have several western species, largely montane, but coastal mitrewort occurs in woods along the coast from central California to British Columbia and may come down to the edge of the shore. This species blooms largely in April and May.

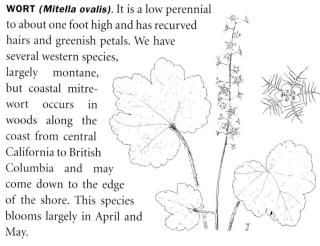

ALUMROOT, or **SMALL-FLOWERED HEUCHERA**, *(Heuchera micrantha)* is another member of the saxifrage family. It grows near the coast from San Luis Obispo County, California, to southern Oregon on rocky banks and in humus. It has a well-developed woody base, basal leaves with five to seven lobes, and stoutish flowering stems one to two feet high. The whitish flowers are minute but borne in great profusion and appear from May to July.

Another species of the immediate coast is **SEASIDE HEUCHERA,** or **SEASIDE ALUMROOT *(Heuchera pilosissima),*** a robust plant with flowers more rounded at the base, pinkish white, and with shorter styles. It is a perennial that grows from an elongate rootstock and has flowering stems one to two feet high and rounded basal  leaves one to three inches across. The inflorescence is rather narrow and compact, and the petals are pinkish white and very small. The species occurs on wooded slopes below 1,000 feet from San Luis Obispo County, California, to Humboldt County and flowers from April to June.

In the gooseberry family (Grossulariaceae) are the currants and gooseberries (*Ribes* spp.). **CANYON GOOSEBERRY *(Ribes menziesii)*** has several forms, one of which reaches the coast. It is loosely branched, spiny, shrubby, three to six feet tall, bristly, and hairy. The rather firm leaves are one-half to one-and-a-half inches across and have gland-tipped hairs beneath. The flowers have white petals and purplish sepals and produce a globular, bristly berry. The range is from southern Oregon to south-central California.

Canyon gooseberry

Modesty, or yerba de selva

In the mock-orange family (Philadel-phaceae) is a trailing, slightly woody plant, **MODESTY,** or **YERBA DE SELVA,** *(Whipplea modesta).* The branches are weak and slender, and the leaves are op-posite and deciduous. The small, white

J

flowers are crowded into terminal clusters with five to six thin, erect sepals and five to six white, spreading petals. The plant is named for Lieutenant Whipple, commander of a government exploring expedition to Los Angeles in 1853 and 1854. It ranges in shaded places in the Coast Ranges from Monterey County, California, to Oregon.

The rose family (Rosaceae) is near the saxifrage family (Saxifragaceae) in having a sort of tube at the base of the flower, with the sepals and petals arising from the rim. A mostly herbaceous genus of this family is *Horkelia,* which has white

California horkelia

flowers having 10 stamens with dilated filaments, as in **CALIFORNIA HORKELIA *(Horkelia californica)***. It is a glandular perennial and rather pleasantly aromatic. The main leaves are largely basal and four to eight inches long and have five to eight or more pairs of leaflets. The flower tube is cup shaped, and the white petals are about one-fourth inch long. It is found in grassy places near the coast of central and northern California.

A beach dweller in the rose family is **BEACH STRAWBERRY *(Fragaria chiloensis)***. It spreads by runners, or stolons, and has three leaflets that are shiny above and silky beneath. The

Beach strawberry

pure white flowers are staminate or pistillate and are usually produced on different plants. Occurring on beaches and adjacent bluffs, this strawberry ranges from central California to Alaska and is found also in Hawaii and South America. The Chilean form is one of the ancestors of domestic strawberries.

Also in the rose family is **CALIFOR-NIA ACAENA** *(Acaena pinnatifida var. californica),* a rather remarkable plant in that the flower tube is armed with recurved barbed prickles. It is a perennial herb and one-third to two feet high and has deeply cut leaflets that are silky beneath. The sepals are green, there are no petals, and the stamens are dark purple. It is a plant of the coastal strand and adjacent  bluffs from Sonoma to Santa Barbara Counties, California.

California blackberry

One of the large groups in the rose family is the blackberry or bramble complex, which includes **CALIFORNIA BLACKBERRY (Rubus ursinus)**. It is a green mound builder and a trailer or partial climber and has long stems that root at the tips and have many straightish bristles. It is a variable plant, and one form near the coast has leaves that are bright green above and, at most, lightly hairy beneath, in contrast to another form that has duller leaves that are more or less feltlike beneath. Flowers are white in both forms and produce black berries if not in too dry a place. California blackberry grows along much of the California coast, and the berries are delicious eaten fresh or made into tasty preserves and pies. The cultivated berries like youngberry, boysenberry, and ollalieberry have been developed from this species.

Thimbleberry

A different sort of *Rubus* is **THIMBLEBERRY** *(Rubus parviflorus)*. It is deciduous and without prickles and has shreddy bark in age. The leaves are lobed, rather than divided into separate leaflets, and are four to six inches wide. The scarlet, hemispheric fruit is about half-an-inch in diameter. The species occurs widely, but along the coast it may be found near woods and thickets from Santa Barbara County, California, to Alaska. It flowers from March to August and produces rather flavorless fruit.

An airy, spiraea-like shrub, **OCEANSPRAY,** or **CREAM BUSH,** *(Holodiscus discolor)* is also a member of the rose family. It is a spreading shrub and four to 18 feet high. A number of named forms vary in leaf size and teeth but in general grow in rocky places such as sea bluffs and canyons on and away from the immediate coast, from British Columbia to southern California. The small flowers are whitish and in large clusters and appear from May to August.

Oceanspray, or cream bush

Oregon crab apple

One group of the rose family has applelike, or pome, fruits with persistent sepal tips at the end. **OREGON CRAB APPLE (Malus fusca)** is a large shrub or small tree with white flowers an inch in diameter and oblong, purple black fruits half-an-inch long. It is native along the north coast from Sonoma and Napa Counties, California, to Alaska. It flowers from April to June.

Another pome fruit is **PACIFIC SERVICE-BERRY (Amelanchier alnifolia var. semiintegrifolia)**. It is a tall, slender shrub with erect branches and oblong to rounded leaves over an inch long. The fragrant flowers are in small, erect clusters and have white petals. The fruit is purplish black when ripe and half-an-inch in diameter. This species grows along the coast in moist and open places from northern California to Alaska. Its flowers appear from March to May.

Oso berry

Catalina crossosoma

OSO BERRY (*Oemleria cerasiformis*) is a shrub with simple, entire, deciduous leaves and nodding clusters of fragrant flowers. The cherrylike fruit is black with bitter pulp. It is not confined to the coast but does occur there in canyons and similar places from Santa Barbara County, California, to British Columbia.

A very interesting plant botanically is **CATALINA CROSSOSOMA (*Crossosoma californicum*),** in the small crossosoma family (Crossosomataceae). It has characteristics of both the much larger buttercup family (Ranunculaceae), such as several free pistils, and of the rose family (Rosaceae), such as a more highly developed floral tube. The species shown has pure white flowers in early spring and occurs only on Santa Catalina, San Clemente, and Guadalupe Islands. A related species of the deserts is smaller and less conspicuous.

The pea family (Fabaceae) is one of our two largest families and has many thousands of species worldwide. Among the larger genera is *Astragalus,* consisting of rattleweeds (so called because their seeds rattle around in the inflated seedpods) and locoweeds (so called because some species can poison livestock and thus craze the animals). **MENZIES' RATTLEWEED (*Astragalus nuttallii*)** is a robust perennial, becoming one to almost three feet tall, although it can be low and matted in windy locations. The flowers are greenish white and about half-an-inch long, and the bladdery pods are one to over two inches long. The species occurs on the mainland coastal strand from Monterey Bay, California, to Point Conception, Santa Barbara County.

SOUTHERN CALIFORNIA LOCO-WEED *(Astragalus trichopodus* var. *lonchus),* differs from Menzies rattleweed *(A. nuttallii)* because its seedpods are on little stalks, or stipes, above the calyx. The flowers are somewhat larger, and the leaflets are not notched at the tips. It grows on sandy bluffs and low hills along the immediate coast or sometimes on shingle banks behind the barrier beaches. It ranges from Ventura County, California, to northern Baja California, occurring also on several of the Channel Islands.

Southern California locoweed

We have several species of pea (*Lathyrus* spp.) on the West Coast. One, **SILKY BEACH PEA** *(Lathyrus littoralis),* is a white-silky perennial with four to eight leaflets. The flowers are mostly in groups of two to six, vary from white to pink to purple, and measure half-an-inch or longer. This species occurs on the coastal strand from Monterey County, California, to British Columbia and bears flowers from April to July.

Silky beach pea

Redwood-sorrel

The oxalis family (Oxalidaceae) is not a family of sea beaches, at least in our part of the world, but **REDWOOD-SORREL** *(Oxalis oregana)* is a plant of woods close to the shore and may sometimes be found very near it. It has wiry, scaly, branching rootstocks with tufts of cloverlike leaves. The white to deep pink flowers, which are often veined purple, have five sepals and petals and 10 stamens, five long and five short. This plant ranges from Monterey County, California, to Washington.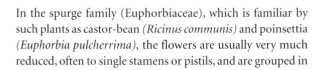

In the spurge family (Euphorbiaceae), which is familiar by such plants as castor-bean *(Ricinus communis)* and poinsettia *(Euphorbia pulcherrima),* the flowers are usually very much reduced, often to single stamens or pistils, and are grouped in clusters. The pistil has a three-cornered ovary that may become quite conspicuous. **CALIFORNIA CROTON (*Croton californicus*)** is a more or less hoary perennial one to three feet high and has leaves one-half to one-and-a-half inches long. It has no petals and many stamens, and the pistils form capsules one-fourth inch in length. Cali-

fornia croton is found in the interior and also on the beaches and coastal bluffs from the San Francisco Bay Area to Baja California.

An important northwestern shrub is **CASCARA (*Rhamnus purshiana*).** It belongs to the buckthorn family (Rhamnaceae), together with California-lilacs (*Ceanothus* spp.). Again, it is

Cascara

not a beach plant but occurs in nearby forests; however, it can be found along the shore, especially in brushy clearings. It is a deciduous shrub or small tree and has smooth, green leaves two to six inches long. The small flowers are followed by rather persistent roundish black berries. The inner bark of this species was once an important source of a well-known laxative. The geographical range is from northern California to British Columbia and Montana.

In the carrot family (Apiaceae) are some small aquatic plants such as **WESTERN LILAEOPSIS** *(Lilaeopsis occidentalis),* a low, tufted, creeping perennial from long rhizomes. The leaves have no flattened blades but are reduced to linear structures with transverse partitions. The small clusters of tiny white flowers are an inch or so tall and produce small, rounded, somewhat corky fruits. Western lilaeopsis is found in and near coastal salt marshes from Solano and Marin counties, California, to British Columbia. It bears its flowers from June to August.

Pacific oenanthe, or American oenanthe

A larger umbellifer (member of the carrot family) of wet places is **PACIFIC OENANTHE,** or **AMERICAN OENANTHE,** *(Oenanthe sarmentosa),* a succulent-stemmed perennial to about four feet long, with much divided leaves four to 12 inches long. The small, white flowers are in large umbels and have five sepals and five petals. The ovary produces an oblong, often purplish fruit about one-eighth inch long. It is found in marshes and sluggish water along and away from the coast, from California to British Columbia and Idaho, and bears flowers from June to October.

Likewise in the carrot family is **COAST ANGELICA** *(Angelica hendersonii)*. It is a stout perennial and one to almost three feet high and has large leaves that are green above and white woolly

Coast angelica

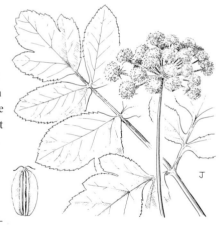

beneath. The small, white petals are woolly on the back, and the somewhat winged fruit, shown at the lower left of the illustration, is about one-third inch long. Found along the coast, both on the coastal strand and neighboring bluffs, from central California to southern Washington, it blooms in June and July.

Another plant with compound umbels, that is, small umbels on stalks radiating out from one level, is **HEMLOCK-PARSLEY** *(Conioselinum pacificum)*. It is perennial, stout, mostly branched, glabrous, and from one to five feet tall. The leaves are two to eight inches long, on equally long or longer stems.

The white flowers are practically without sepals and have notched petals. The fruit, shown at the lower right of the drawing, is about one-sixth inch long. This species inhabits ocean bluffs, cold marshes, and the like and ranges from northern California to Alaska and is also found in Siberia and our Atlantic Coast.

POISON-HEMLOCK *(Conium maculatum)* is a tall biennial herb with spotted stems and multicompounded leaves. The small white flowers are arranged in many-rayed compound umbels. It is naturalized from Europe and has established itself in low places and in great masses along parts of our Pacific Coast. The flowers appear from April to July.

Poison-hemlock

COW-PARSNIP *(Heracleum lanatum)* is a common plant in the mountains below 9,500 feet but is also well represented along the coast from Monterey County, California, to Alaska. Perennial, three to eight feet high, and somewhat woolly, it has large, rounded leaves four to 20 inches broad and more or less lobed. The white flowers are in large, flat umbels on stems with conspicuous expanded bracts, or modified leaves. Flow-

Cow-parsnip

ering is from April to July, and it produces flattened fruits one-third to one-half inch long.

The last member of the carrot family described in this book is **AMERICAN GLEHNIA** *(Glehnia littoralis* subsp. *leiocarpa),* an almost stemless, more or less fleshy perennial. The sheathing

American glehnia

leaf stems are fairly well buried in the sand, and the leaf blades are one to six inches long and woolly beneath. The rest of the plant is also quite hairy. It has a typical umbelliferous flower with five sepals, five petals, and an inferior ovary and winged fruit. American glehnia grows in beach sand from Mendocino County, California, to Alaska and flowers in May and June.

The manzanitas (*Arctostaphylos* spp.), of the heath family (Ericaceae) are among the most important woody genera of the West. Manzanitas are generally characterized by their leathery leaves, usually reddish bark, small, urn-shaped flowers, and small, usually reddish fruits. A species of acidic, often moist places along the coast is **FORT BRAGG MANZANITA (*Arctostaphylos nummularia*)**. It varies from almost prostrate to erect, and the flowers have four sepals and four corolla lobes. It also has a southern form with larger leaves, and the two forms range from the Santa Cruz Mountains, California, to Mendocino County.

Most manzanitas have five-parted flowers, as does the coastal form of the transcontinental **BEARBERRY,** or **SANDBERRY, (*Arctostaphylos uva-ursi*)**. It is a prostrate plant with trailing stems that send up erect branches a few inches high. The white to pinkish corolla is one-sixth inch long, and the red berries are somewhat larger. It is found in sandy places along the coast from central California to Alaska and east. The more common man-

Fort Bragg manzanita

zanitas are shrubbier, often quite large, and grow in dry, often rocky places well up into the mountains.

"Manzanita" means little apple, and the applelike character of the small, reddish fruits is evident in **HAIRY MANZANITA** **(Arctostaphylos columbiana).** This species is well distrib-

Hairy manzanita

uted, much branched, and up to eight feet tall and has pale gray leaves one to more than two inches long. It is found in rocky places from British Columbia to Sonoma County, California.

Also in the heath family is **CALIFORNIA HUCKLEBERRY** *(Vaccinium ovatum),* a stout, much-branched, evergreen shrub. Its smooth, shining leaves are one-half to one-and-a-half inches long and somewhat toothed on the edges. The bell-shaped flowers are white to pink, about one-fourth inch long, and produce sweet black berries edible to humans and one-third inch long. Not a beach plant, it is usually found in woods but comes out to the coast in disturbed places and ranges from British Columbia to central California. It flowers from March to May or June.

Another member of the heath family is **SALAL** *(Gaultheria shallon).* It is a spreading shrub or subshrub one to five feet

California huckleberry

Salal

tall and has tough evergreen leaves one to four inches long. The urn-shaped flowers are white to pink and one-third inch long and are arranged in clusters up to six inches long. The fruit is a dark purple capsule. Salal grows in brushy or wooded places near the coast from Santa Barbara County, California, to British Columbia and flowers from April to July. Its leathery leaves are used in large quantities by florists, who have named them "lemon leaves."

WESTERN AZALEA (*Rhododendron occidentale*) is also in the heath family. It is a deciduous shrub, loosely branched, and three to 12 feet tall and has shredding bark and thin, light green leaves one to three inches long. It grows in moist places, coming out to the coast from Umpqua Valley, Oregon, southward to Santa Cruz County, California. Flower color varies from rose, to whitish with a central salmon spot, to pink with an orange flush. Flowering season is April to June and later in the mountains.

Western azalea

The so-called **SEA-MILKWORT**, or **GLAUX**, *(Glaux maritima)* is a fleshy perennial. It is in the primrose family (Primulaceae), which includes plants such as primroses (*Primula* spp.), cyclamen (*Cyclamen* spp.), and shooting stars (*Dodecatheon* spp.). This species has a bell-shaped calyx about one-eighth inch long, no petals, and five stamens. The stems are usually less than a foot tall, and the leaves are about half-an-inch long. It is found in coastal salt marsh and on coastal strand from San Luis Obispo County, California, to Alaska, as well as on the Atlantic Coast and in Europe.

Another plant of the primrose family is **PACIFIC STARFLOWER** *(Trientalis latifolia)*, a neat, slender little plant

Pacific starflower

that is two to eight inches high. It grows from an underground tuberous rootstock, and has a whorl of four to six leaves near the top. The pinkish white flowers are about half-an-inch wide and have five to seven sepals and petals. The flowers appear from April to July. Pacific starflower grows in shaded places, chiefly in woods, and may be found adjacent to the shore from San Luis Obispo County, California, to British Columbia.

The genus *Plantago* of the plantain family (Plantaginaceae) is quite cosmopolitan. In America we have a number of bad weeds belonging to this genus, introduced mostly from Europe, but we also have a good many native species, such as **PACIFIC SEASIDE PLANTAIN** *(Plantago maritima)*. This is a small perennial with many strongly ascending narrow leaves and a few somewhat longer, leafless stems bearing spikes of small four-petaled flowers. This plant grows in salt marshes and on the strand, extending from Santa Barbara County, California, to Alaska. It flowers from May to September.

Pacific seaside plantain

An inhabitant of saline and alkaline places, including the coast, is **ALKALI WEED (*Cressa truxillensis*),** a low, much-branched, gray perennial. It belongs to the morning glory family (Convolvulaceae). It is tufted, woolly hairy, and four to eight inches tall and has small white flowers solitary in the upper leaf axils. The corolla is about one-fourth inch long and has five spreading lobes. It ranges from Oregon to Mexico and Texas and blooms from May to October.

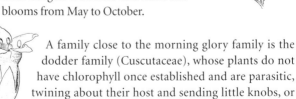

A family close to the morning glory family is the dodder family (Cuscutaceae), whose plants do not have chlorophyll once established and are parasitic, twining about their host and sending little knobs, or

Alkali weed

Canyon dodder

haustoria, into it in order to obtain their nourishment. Leafless, or nearly so, they are orange to yellow in color and have small, white flowers. **CANYON DODDER** *(Cuscuta subinclusa)* has rather coarse orange stems, and the calyx is shorter than the corolla tube. It is parasitic on many different plants and ranges from Oregon to Baja California. With more slender stems and a calyx not shorter than the corolla tube is salt marsh dodder *(C. salina),* which grows on *Cressa, Salicornia,* and *Chenopodium* and ranges as far north as British Columbia.

Large-flower linanthus

In the phlox family (Polemoniaceae), we find no true beach plants, but some such as **LARGE-FLOWER LINANTHUS (Linanthus grandiflorus)** are found on the coastal strand and sea bluffs. This plant is annual, erect, and up to one-and-a-half feet high, and its leaves are cleft to the base into five to 11 linear lobes. The flowers are white to pale lilac and up to an inch long. It is primarily a species of central California and blooms from April to July.

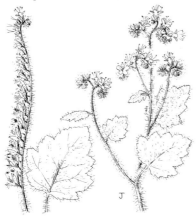

The waterleaf family (Hydrophyllaceae), with its flowers in coiling branches and the ovary forming a capsule, is abundant in western North America. One of its largest constituents is phacelia, or wild-heliotrope, (Phacelia spp.), as the blue-flowered species

Stinging phacelia

are often called. **STINGING PHACELIA (Phacelia malvifolia)**
gets its common name from its stiff, bristly hairs. It is an erect
annual, one to three feet tall, with broad leaf blades to about
four inches long. The dull white flowers are about half-an-
inch across. It is found mostly in sandy and gravelly places,
hence on back beaches, from Oregon to central California
and blooms from April to July.

COMMON PHACELIA (Phacelia distans) is an annual that is six
to 30 inches high, usually branched above, and somewhat stiff
hairy, and its leaves have toothed to deeply divided divisions.
The blue or white, broadly bell-shaped flowers are in coiled
branches and one-fourth to one-third inch long. This species
is common in much of California and grows in sandy coastal
areas from Mendocino County southward. Flowers appear in
spring and early summer.

An inconspicuous group in the waterleaf family is romanzoffia, with three species on the West Coast. **SUKSDORF'S RO-
MANZOFFIA,** or **CALIFORNIA ROMANZOFFIA, (Romanzoffia**

Common phacelia

Suksdorf's romanzoffia, or California romanzoffia

californica) is a perennial with basal, woolly tubers, slender stems up to one foot high, and white funnelform corollas to about half-an-inch long. It and the other species are much alike and occur on ocean bluffs and in moist spots in the rocks, from central California to Alaska. Flowering is largely in spring and summer.

Seaside heliotrope

The borage family (Boraginaceae) resembles the waterleaf family in its coiled inflorescences, or cymes, and its flower shape, but the ovary forms one-seeded nutlets rather than a several-seeded capsule. One of its members is **SEASIDE HE-LIOTROPE** *(Heliotropium curassavicum),* a perennial with underground rootstocks that send up scattered shoots four to 20 inches high. It is glabrous and waxy and has entire, succulent leaves and flowers one-eighth to one-fourth inch wide that are white with yellow spots. The plant is characteristic of saline spots; hence, it is not surprising to find it growing on the coast, especially in California.

WHITE FORGET-ME-NOT, or **COMMON CRYPTANTHA,** *(Cryptantha intermedia)* is an annual member of the borage family, usually stiff hairy, and six to 18 inches high and has white flowers one-eighth to one-fourth inch broad. The nutlets in each flower are usually four in number and one-seeded. Common in the interior, this species grows along the sea bluffs and on back beaches of southern California and adjacent Baja California. It blooms from March to July.

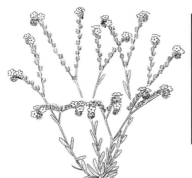

White forget-me-not, or common cryptantha

The nightshade family (Solanaceae) is well known for potato (*Solanum tuberosum*), tomato (*Lycopersicon* spp.), petunia (*Petunia* spp.), and many other common plants. Tobacco is also in this family, and among the western species is **CLEVELAND'S TOBACCO (*Nicotiana clevelandii*)**, a glandular-hairy annual that is one to two feet tall and much branched and has leaves to three inches long. The greenish white flowers have a rather long corolla tube and are more than half-an-inch across. It is occasional in sandy places on the coastal strand and adjacent bluffs and ranges from Santa Barbara County, California, southward.

Everyone who visits or lives on the Pacific Coast should learn to recognize the native shrub **POISON-OAK (*Toxicodendron diversilobum*)**—but at a distance. It is widespread in many plant communities, growing prostrate along the ground, erect as a shrub, or climbing in trees via adhesive rootlets. Despite its common name, it is not an oak but belongs to the mostly tropical sumac, or cashew, family (Anacardiaceae). Humans may develop a persistent itching, oozing dermatitis even after slightly brushing its leaves, from inhaling its smoke if burning, or from petting dogs that have frolicked in its thickets. Immune individuals may become highly sensitized at any age. Despite its noxious qualities, its small, whitish flowers are pleasantly perfumed, its white berries are ornamental after the leaves have fallen, and its green usually glossy leaves are striking. As early as July the leaves turn brilliant shades of red, yellow, and bronze. But remember: leaflets three, let it be!

Poison-oak

Most people know elderberry (*Sambucus* spp.), especially the blue-berried forms, but a red-berried species found near the coast is **RED ELDERBERRY (*Sambucus racemosa* var.**

Red elderberry

racemosa). A member of the honeysuckle family (Caprifoli-aceae), red elderberry is a shrub six to 18 feet high and has leaves composed of five to seven large leaflets. The inflores-cence is two to four inches across and has many small, whitish five-lobed flowers. The bright scarlet fruits are about one-sixth inch in diameter. Found on flats and coastal bluffs, the species ranges from British Columbia to central California. Elderberries have occasionally poisoned humans, and it is best to avoid eating the berries of any species.

Plants in the sunflower family (Asteraceae) have many minute flowers, or florets, packed into a head that is surrounded below by an involucre of overlapping bracts. One of these plants is **POVERTY WEED** *(Iva axillaris* **subsp.** *robustior),* a low herb

Poverty weed

that spreads from slender rhizomes and has leafy stems. The involucre is cup shaped and surrounds the inconspicuous greenish white flowers, of which the marginal are fertile and produce seed, whereas the inner are staminate only. It is found in coastal salt marsh, at least in southern California, and east of the mountain ranges as far north as Washington.

COYOTE BRUSH, or **CHAPARRAL BROOM,** *(Baccharis pilularis)* ranges from San Diego County, California, along the coast northward to Tillamook County, Oregon. In central California it also grows in the Sierra Nevada foothills. This evergreen shrub often forms pure stands and is able to toler-

Coyote brush, or chaparral broom

ate poor soils and windswept conditions. Although it is hardly a dazzling ornamental, its toughness has led a prostrate coastal race to be widely planted as a durable ground cover. Coyote brush is unusual for a member of the sunflower family because it has separate sexes on separate plants. It begins to flower in August, and the male plants can be distinguished from females because the flower heads of the two sexes are visibly different. When the female plants go to seed, they become covered with white down from their wind-dispersed seeds and resemble giant cauliflowers.

Another representative of the sunflower family, **PEARLY EVERLASTING *(Anaphalis margaritacea)*,** is not primarily a beach plant but in the north comes down to the very edge of the sand. A white-woolly perennial with slender running rootstocks, it forms patches in which the erect stems are commonly one to two-and-a-half feet high. The leaves are narrow, stemless, and one to four inches long. The several papery, pearly white bracts surrounding the flower head are a conspicuous feature of this plant. In the coastal regions it ranges from Monterey County, California, to Alaska but reaches farther south in the mountains.

Pearly everlasting

Silver beachweed, or beach-bur

Another member of the sunflower family, but with smallish heads of inconspicuous flowers, is **SILVER BEACHWEED**, or **BEACH-BUR**, *(Ambrosia chamissonis)*. A silver-haired perennial with radiating procumbent (prostrate below, erect above) stems, it forms loose mats on the sand. The variable leaves can be barely toothed to deeply lobed. The stems bear terminal spikes with heads of staminate flowers above and spiny, solitary pistillate heads below that become burlike in fruit. The range is from British Columbia to Baja California.

Several white-woolly, not very woody species of *Artemisia* are found on sandy shores of large lakes and seas of much of the northern hemisphere. Such a species is **COASTAL SAGEWORT**

(Artemisia pycnocephala), which is more or less woody at the base, mostly one to two feet tall, densely leafy, and whitish or grayish silky-woolly and has leaves dissected into linear lobes. The many small heads are borne in dense clusters on long branches. This is a plant of the beaches from Monterey County, California, to southern Oregon. Farther north it is replaced by similar forms of Pacific sagewort (*A. campestris* subsp. *borealis*), with more appressed and silky pubescence.

YARROW, or **MILFOIL,** *(Achillea millefolium)* is an aromatic perennial herb with finely dissected leaves and many small heads of flowers arranged in flat-topped inflorescences. Along the coast is a form whose stems are one to two feet tall and whose leaf segments are very numerous and thick. An inhabitant of the coastal strand from Santa Barbara to Del Norte Counties, California, this plant blooms in June and July. Farther north, other forms with the same general appearance may grow near the coast.

In the chrysanthemum group is **OX-EYE DAISY *(Leucanthemum vulgare)*,** which has white, petal-like ray flowers and yellow, tubular disk flowers. It is a perennial, with solitary heads one to two inches across. It is a native of Europe and for a long

Ox-eye daisy

time has been naturalized in the eastern United States. More recently and increasingly, it has established itself in the Pacific states, especially northward. The leaves are for the most part not divided, and the plant is almost hairless. Flowering is largely in late spring and early summer.

Members of the *Malacothrix* genus, also in the sunflower family, have strap-shaped ray flowers and no tubular disk flowers. A coastal species is the white-headed **CLIFF MALACOTHRIX (Malacothrix saxatilis)**, a plant of sea bluffs. A narrow-leaved form, *M. saxatilis* var. *tenuifolia,* grows on coastal strand as well as bluffs. A form found mostly on the islands off southern California, *M. saxatilis* var. *implicata,* has leaves divided into linear segments. The ray flowers are about two-thirds of an inch long and may be rose or purplish. All forms bloom much of the year.

Cliff malacothrix

MONTEREY PINE (Pinus ra-diata) is one of the well-known coastal trees of California, although it grows naturally in only a few spots in Santa Cruz, Monterey, and San Luis Obispo Counties. It is in the pine family (Pinaceae) and can be distinguished by having rather dark green foliage, mostly with three needles in a cluster, and asymmetrical cones that remain closed and attached to the branches for many years. The wood is light, soft, close grained, and not strong. A similar closed-cone pine is bishop pine *(P. muricata)*, but it is mostly two needled, the needles being four to six inches long. It grows from Humboldt to Santa Barbara Counties, California.

Another coastal pine is **SHORE PINE (Pinus contorta subsp. contorta)**, which has clusters of mostly two needles up to about two inches long and almost symmetrical, open cones that are deciduous when mature. It is found on coastal strand and adjacent bluffs from Mendocino County, California, to Alaska and has a light, hard, strong, brittle, coarse-grained wood occasionally used for fuel.

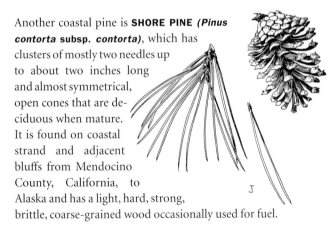

TORREY PINE (Pinus torreyana) is quite different from the Monterey pine *(P. radiata)* and shore pine *(P. contorta* subsp. *contorta)* in having five needles in a cluster that are eight to 12

inches long and gray green. The cones too are large: four to six inches long. Like Monterey pine, Torrey pine now has a very restricted range as compared to that of prehistoric times, being found only in the region around Del Mar in San Diego County, California, and on Santa Rosa Island. It has a light, soft, coarse-grained wood and, like Monterey pine, is cultivated in New Zealand, Kenya, and other warmer regions. Like other coastal pines it takes on interesting shapes along the windswept coast.

Torrey pine

Another coastal conifer of the pine family is **SITKA SPRUCE (Picea sitchensis)**. Spruces bear their short needles singly, not clustered, and the branchlets are roughened by the persistent needle bases. The needles are sessile, usually more or less four-sided, and often sharp pointed. The cones of this species are

Sitka spruce

oblong, two to four inches long, and the papery, scalelike bractlets below the cone scales are hidden. This is a tree of the coastal strand and adjacent areas, from Mendocino County, California, northward to Alaska.

DOUGLAS-FIR (Pseudotsuga men-ziesii), another member of the pine family, is often confused with spruce (*Picea* spp.), but its branch-lets are not roughened and the bracts of the cone scales are conspicuous. The needles are flat in cross section and ap-pear two ranked. It is not primarily a coastal tree but often comes down to the coast. It occurs from Monterey Bay, California, north-ward to British Columbia and into the Rocky Mountains. It is the most important lumber tree of North

America, and its wood is known in the trade as "Oregon pine." It is perhaps the most commonly used Christmas tree.

WESTERN HEMLOCK
(Tsuga heterophylla) is the last member of the pine family mentioned here. Like spruce species (*Picea* spp.), it has branchlets roughened by persistent needle bases. The branches are slender, more or less pendulous, and bear flat, more or less two-ranked leaves one-fourth to three-fourths of an inch long. The cones are up to about one inch long and have rather thin, persistent scales. Hemlock bark has been important in tanning, and the rather durable wood of this

Western hemlock

species is used for construction. It is another forest tree that sometimes reaches the coast, where it may be found from northern California to Alaska.

REDWOOD *(Sequoia semper-virens)* is, of course, one of the famous trees of the world because of its great height and majesty. A member of the bald-cypress family (Taxodiaceae), it has very durable and straight-grained wood. The red, spongy-fibrous bark is conspicuous. The leaves are linear and one-half to one inch long and spread in two ranks.  Redwood forest is a distinct plant community in northwestern California, inhabiting the coastal fog belt and coming out to the shore at the mouth of canyons and gulches. Redwood ranges from the Santa Lucia Mountains of south-central California to southwestern Oregon.

Redwood

A tree that has been widely introduced into cultivation and is picturesque on the coast is **MONTEREY CYPRESS (*Cupressus macrocarpa*),** of the cypress family (Cupressaceae). The leaves are scalelike, about one-twelfth inch long, and bright green. The persistent cones are globose or slightly elongate, an inch or 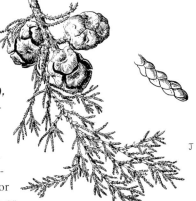 more long, and remain closed for many years. The eight to 12 cone scales bear many seeds. Monterey cypress is practically confined as a native to the Monterey Peninsula of California. In cultivation, especially in warmer areas away from the coast, it is unfortunately subject to a fungus disease.

Monterey cypress

The last conifer I will mention is **WESTERN RED-CEDAR,** or **CANOE-CEDAR**, *(Thuja plicata),* also in the cypress family. A forest tree buttressed at the base, it inhabits moist places in the outer Coast Ranges from Mendocino County, California, to Alaska and extends inland to Montana. The leaves are scalelike and mostly about one-eighth inch long, and the cones are about half-an-inch long and made up of eight to 12 scales. The wood is light and soft, easily split, so it is used for shingles and also for interior finishes.

RED ALDER, or **OREGON ALDER,** *(Alnus rubra)* is in the birch family (Betulaceae). The flowers are in catkins; this species has deciduous staminate catkins (upper right in the illustration) and more persistent, conelike pistillate catkins (upper left). It is a deciduous tree characterized by few-scaled, long, winter leaf buds and by toothed leaves. Red alder differs from the inland species, white alder *(A. rhombifolia),* by having leaves that are rusty haired beneath and slightly inrolled on the edges, and by its narrowly winged seeds. The range is in damp places from Santa Cruz County, California, to Alaska.

In the beech family (Fagaceae), along with oak (*Quercus* spp.) and chestnut (*Castanea* spp.), is **GIANT CHINQUAPIN (*Chrysolepis chrysophylla*)**. It is a tall tree with heavily furrowed bark, leathery leaves two to six inches long, and staminate catkins (as shown in the upper right of the il-lustration). The pistillate structures produce burs (lower left in the illustration) with long spines and enclose one to three nuts that are up to about half-an-inch long. The fruit matures in the second season. This is a forest tree coming out to the beach and ranging from Mendocino County, California, to Washington.

Like giant chinquapin (*Chrysolepis chrysophylla*) because it has long, staminate, odorous catkins and like the oak (*Quercus* spp.) because it has an acorn is **TANBARK-OAK,** or **TAN-OAK, (*Lithocarpus densiflorus*),** also in the beech family. This tree is evergreen with a narrow, conical crown. The acorn cup has slender, spreading scales and is quite different from that of the oak. In North America we have a single species throughout the Coast Ranges from Ventura County, California, to southern Oregon, but southeastern Asia has about 100 species. In the mountains our species has a dwarf, shrubby form.

Coast live oak, or encina

Oaks (*Quercus* spp.), have a tremendous number of species, and one of the most common in California is **COAST LIVE OAK**, or **ENCINA** *(Quercus agrifolia)*. It is a member of the beech family and a broad-headed tree that takes on very picturesque shapes in age. The rather harsh leaf blades are characteristically inrolled slightly at the edge, and the acorns are long and pointed. Common over much of California between the Sierra Nevada and more southern mountains and the coast, the species reaches the shore itself, especially at the mouth of water courses. In much of central California it is often festooned with a netlike gray lichen sometimes incorrectly called "Spanish-moss."

CALIFORNIA BAY, or **CALIFORNIA LAUREL,** *(Umbellularia californica)* is another widely distributed tree in California and extreme southwestern Oregon, where it is called **OREGON MYRTLE**. It is a member of the laurel family (Lauraceae). It is pungently aromatic, has stiffish or rubbery leaves and small, yellow green flowers. The leaves are often used for seasoning as a substitute for the true bay (*Laurus* spp.). The wood is hard, strong, and takes a high polish, so it is very good for turning. Like the other trees described in this section, it is primarily a forest tree but often reaches the shore. It blooms from December to May.

Vine maple

Maples (*Acer* spp.) have opposite, broad, usually lobed but sometimes compound leaves and are in the maple family (Aceraceae). The small flowers are variously clustered and have four to nine stamens and a two-styled pistil that forms a two-winged fruit, or samara, united below. The species most apt to appear on the shore is **VINE MAPLE** *(Acer circinatum)*. It is vinelike and reclining, with slender twigs and leaves two to five inches across. This species ranges from northern California to British Columbia. Big-leaf maple *(A. macrophyllum)* is stouter and has much larger leaves.

GLOSSARY

Achene A dry, nonopening, one-seeded fruit.

Anther The pollen-forming portion of a stamen.

Axil The upper angle of a leaf and branch.

Banner The uppermost, often largest petal of flowers (as in the pea family).

Calyx The usually green, outer series of flower leaves or sepals.

Catkin A spike of unisexual flowers.

Corolla The part of the flower consisting of the petals; collective term for petals.

Cotyledon A seed leaf or first leaf of an embryo.

Cyme A branched inflorescence in which the central flower blooms first.

Decumbent (plant) A plant that mostly reclines on the ground but with stems or flowers that curve upward.

Dicotyledon The larger main group of flowering plants. Dicotyledons usually have two cotyledons.

Filament The stalk of an anther.

Floret A single flower with immediately subtending bracts (in the grass and sunflower families).

Frond A leaf (often referring to ferns).

Funnelform (corolla) A corolla that widens from the base into an ascending, spreading shape.

Glabrous (plant, etc.) A plant or plant structure that is devoid of hairs.

Glaucous (plant, etc.) A plant or plant structure that is covered with a white, waxy bloom.

Globose (plant structure) A spherical plant structure.

Glume One of the usually paired bracts at the base of a grass spikelet.

Halophyte A salt-tolerant plant.

Haustorium A sucker of a parasitic plant. The plural is *haustoria*.

Herbaceous Like an herb, not woody; or, having a green color and a leafy texture.

Inferior (ovary) An ovary positioned below the insertion of other flower parts (corolla, calyx).

Inflorescence An entire cluster of flowers and associated structures.

Involucre A group of bracts held together as a unit usually forming a cuplike structure.

Keel The two lowermost fused petals of many members of the pea family.

Margin The edge, usually of a leaf or flower part.

Mesophytic A plant that has intermediate moisture requirements.

Monocotyledon The group of plants having a single cotyledon.

Obovate (shape) An ovate shape with a more broad distal end.

Ovary The ovule-bearing portion of pistil normally developing in fruit as ovules become seeds.

Ovate Egg-shaped.

Ovule The structure within the ovary containing an egg.

Palmate Lobed or veined so that sinuses radiate from a common point (as in a hand).

Palmately compound (leaf) A leaf made up of palmate leaflets.

Parasite A plant that gets its living from another living plant to which it is attached.

Perianth The structure of a flower comprising the calyx and corolla.

Petal An individual member of a corolla.

Petiole A leaf stalk connecting the leaf to the plant stem.

Pinnate Leaflets arranged on each side of a common petiole as in a fern.

Pinnately compound (leaf) A compound leaf made up of pinnate leaflets.

Pistil The female organ of a flower, comprised of the ovary, style, and stigma.

Pistillate A flower that has fertile pistils but sterile or missing stamens.

Pollen The fertilizing dustlike powder produced by anthers.

Procumbent (plant) A plant that leans forward or along the ground.

Puberelent (plant) A plant that has hairs visible only with magnification.

Pubescent Covered with soft, downy hairs.

Salverform (corolla) A corolla that has a slender tube and an abruptly spreading tube or throat.

Samara A winged fruit that remains closed at maturity.

Saprophyte A plant that lives on dead organic matter.

Scape A leafless flower stalk rising from the ground.

Scurfy (plant structure) A scaly plant structure.

Sepal An individual member of the calyx.

Sessile (flower, leaf) A flower or leaf without a petiole, pedicel, peduncle, or other kind of stalk.

Sorus The receptacle of a fern that contains spores. The plural is *sori.*

Spadix A floral spike with a fleshy axis, as in the arum family.

Spathe A large, showy bract enclosing or subtending a flower cluster.

Spikelet A flower or flowers subtended by glumes (in the grass family).

Stamen The male reproductive part of a flower.

Staminate (flower) A flower that has a fertile stamen but sterile or missing pistils.

Stigma The generally terminal part of the pistil on which pollen is normally deposited.

Stipe A leaf stalk of a fern or of a pistil.

Stipule An appendage at the base of leaves in some plants.

Stolon A runner forming roots or erect stems.

Style A stalklike portion that connects the ovary to the stigma in many pistils.

Superior (ovary) An ovary positioned above the point of attachment of other flower parts (corolla, calyx).

Umbel An inflorescence in which three to many pedicels radiate from a common point.

Viscid (plant) A plant that has a sticky surface.

ART CREDITS

Photographs credited to the California Academy of Sciences Collection are also credited to their individual photographers.

Line Illustrations

DICK BEASLEY 7, 176

TOM CRAIG 59 (bottom), 104, 212

RODNEY CROSS 28, 102, 156, 162

PETER GAEDE 42 (top), 44, 48, 53 (middle), 61, 64 (top), 65, 71, 72, 82, 87 (top, bottom), 101 (top, bottom), 107, 141, 143, 150 (bottom), 164 (bottom), 195, 200 (top, bottom)

JEANNE R. JANISH 26, 27 (top), 27 (bottom), 28, 30, 36, 37, 38, 41, 42 (bottom), 46 (top, bottom), 49 (top), 53 (top, bottom), 54 (top, bottom), 57, 59 (top), 62, 64 (bottom), 66, 74 (top, bottom), 75 (top, bottom), 83 (top, bottom), 86 (top, bottom), 88, 96, 99, 103, 108 (top, bottom), 115, 119, 120, 122 (top, bottom), 123, 124, 125, 126, 132, 134, 135, 136, 137 (top, bottom), 139, 142, 144, 146 (top, bottom), 149, 150 (top), 151 (top, bottom), 152, 153, 154 (top, bottom), 155 (top, bottom), 157 (top, bottom), 158, 159 (top, bottom), 160 (top, bottom), 161 (top, bottom), 163, 164 (top), 165 (top, bottom), 166, 167, 168, 171, 173 (top, bottom), 177 (top, bottom), 178, 179, 180, 182 (top, bottom), 186, 188 (top, bottom), 190 (top, bottom), 194 (bottom), 196, 201, 204 (top, bottom), 205, 206 (top, bottom), 207, 208, 209, 210 (top, bottom), 211 (top, bottom), 213

PAULA NELSON AND BILL NELSON 23

STEPHEN TILLETT 40, 51, 84, 169, 175, 194 (top)

MILFORD ZORNES 49 (bottom)

Color Photographs

SHERRY BALLARD 213

BROTHER ALFRED BROUSSEAU 65 (top), 103 (left), 116 (bottom), 118, 162, 163, 171 (right), 174 (top), 189 (bottom)

CALIFORNIA ACADEMY OF SCIENCES 36, 43 (top), 52 (top), 58, 62, 81 (bottom), 82, 84, 103 (right), 114 (bottom), 116 (top), 119 (bottom), 128, 151, 166 (top), 172 (bottom), 181 (top, bottom), 183 (top), 191, 192 (top), 194, 196, 198, 201 (bottom), 206, 207, 213

PHYLLIS M. FABER 156, 158

WILLIAM T. FOLLETTE ii–iii, vi, xii–xiii, 24–25, 26, 28, 29, 32–33, 34 (top, bottom), 35, 37, 38, 39 (top, bottom), 40, 43 (bottom), 44, 45 (top), 48, 50, 52 (bottom), 55 (top), 56 (top, bottom), 60 (top, bottom), 61, 63, 65 (bottom), 66, 67, 68, 69 (top, bottom), 71, 72, 73, 76 (bottom), 77, 78–79, 80, 81 (top), 83, 85, 87, 88, 89 (top, bottom), 90, 91 (top, bottom), 92, 93, 94 (top, bottom), 95, 96, 97 (top, bottom), 98, 99, 101, 102, 104, 106, 107 (left, right), 108, 109 (left, right), 110–111, 112 (top, bottom), 113, 114 (top), 115, 117, 119 (top), 121 (top, bottom), 123, 124, 127 (top, bottom), 129, 130–131, 133 (top, bottom), 134, 135, 136, 137, 138, 139 (top, bottom), 140 (top, bottom), 141, 143 (top), 144, 145, 146, 147, 148, 149 (top), 152, 153, 159, 167, 168, 169, 170, 171 (left), 172 (top), 174 (bottom), 175, 176, 178, 179, 180, 183 (bottom), 184, 185, 186, 188, 189 (top), 190, 193, 195, 199 (top), 202–203, 208, 209, 212

JOHN GAME 41, 45 (bottom), 47, 49, 51, 54, 55 (bottom), 70, 100, 122, 143 (bottom), 149 (bottom), 166 (bottom), 187, 192 (bottom), 197, 199 (bottom), 201 (top), 205

BEATRICE F. HOWITT 103 (right), 119 (bottom), 151, 166 (top), 201 (bottom)

J.E. (JED) AND BONNIE MCCLELLAN 196, 198

DONALD MYRICK 128, 194

JO-ANN ORDANO 81 (bottom)

ROBERT POTTS 52 (top), 76 (top), 206

CHARLES WEBBER 36, 43 (top), 58, 62, 82, 84, 114 (bottom), 116 (top), 172 (bottom), 181 (top, bottom), 183 (top), 191, 192 (top), 207

INDEX

Page references in **boldface** refer to the main discussion of the species.

ABOUT THE AUTHOR
AND EDITORS

Philip A. Munz (1892–1974) of the Rancho Santa Ana Botanical Garden was professor of botany at Pomona College, serving as dean for three years. Phyllis M. Faber is general editor of the California Natural History Guides. Dianne Lake is rare plant committee cochair and unusual plants coordinator for the California Native Plant Society, East Bay Chapter.

Series Design:	Barbara Jellow
Design Enhancements:	Beth Hansen
Design Development:	Jane Tenenbaum
Composition:	Impressions Book and Journal Services, Inc.
Text:	9.5/12 Minion
Display:	ITC Franklin Gothic Book and Demi
Printer and Binder:	Everbest Printing Company

Return to:
University of California Press
Attn: Natural History Editor
2120 Berkeley Way
Berkeley, California 94720

CALIFORNIA NATURAL HISTORY GUIDES

"It's always good to read a new California Natural History Guide; these little books are small enough to fit into a pocket, inexpensive, and authoritative."   —*Sunset*

"A series of excellent pocket books, carefully researched, clearly written, and handsomely illustrated."   —*Los Angeles Times*

The California Natural History Guide series is the state's most authoritative resource for helping outdoor enthusiasts and professionals appreciate the wonderful natural resources of their state. If you would like to receive more information about the series or other books on California natural history, please fill in this card and return it to the University of California Press or register online at www.californianaturalhistory.com.

Name _____

Address _____

City/State/Zip _____

Email _____

Which book did this card come from? _____

_____

Where did you buy this book? _____

_____

What is your profession? _____

UNIVERSITY OF CALIFORNIA PRESS
www.ucpress.edu

Z ?